Interactions: The Keys to Cereal Quality

Edited by

Rob J. Hamer and R. Carl Hoseney

Published by the

American Association of Cereal Chemists, Inc.

St. Paul, Minnesota, USA

This book has been reproduced directly from computer-generated copy submitted in final form to the American Association of Cereal Chemists by the authors. No editing or proofreading has been done by the Association.

Library of Congress Catalog Card Number: 98-72141
International Standard Book Number: 0-913250-99-6

Printed in the United States of America on acid-free paper

American Association of Cereal Chemists
3340 Pilot Knob Road
St. Paul, Minnesota 55121-2097 USA

CONTRIBUTORS

J.L. Andrews, CSIRO Division of Plant Industry, North Ryde NSW 2113, Australia

F. Bekes, CSIRO Division of Plant Industry, North Ryde NSW 2113, Australia

W. Bushuk, Department of Food Science, University of Manitoba, Winnipeg, MB, R3T 2N2, Canada

D.C. Clark, Institute of Food Research, Norwich laboratory, Norwich Research Park, Norwich NR4 7UA, United Kingdom

L. Dubreil, I.N.R.A. Laboratoire de Biochimie et Technologie des Protéines BP1627 44316 Nantes cédex 03, France

A.C. Eliasson, Department of Food Technology, University of Lund, PO Box 124, S-221 00 LUND, Sweden

P.W. Gras, CSIRO Division of Plant Industry, North Ryde NSW 2113, Australia

R.B. Gupta, Centre for Plant Conservation and Genetics, Southern Cross University, Australia

R.J. Hamer, Wageningen Centre for Food Sciences, PO Box 557, 6700 AN Wageningen, The Netherlands

R.C. Hoseney, R&R Research Services Inc. 8831 Quail Lane Manhattan, KS 66502, United States

F. MacRitchie, Kansas State University, Dept. Grain Science & Industry, Shellenberger Hall, Manhattan KS 66506-2201, United States

D. Marion, I.N.R.A. Laboratoire de Biochimie et Technologie des Protéines BP1627 44316 Nantes cédex 03, France

K. R. Preston, Grain Research Laboratory, Canadian Grain Commission, 1404-303 Main Street Winnipeg, MB, R3C 3G8, Canada

J.H. Skerritt, CSIRO Division of Plant Industry Canberra ACT 2600, Australia

P.L. Weegels, Unilever Research Laboratory, Olivier van Noortlaan 120, 3133 AT Vlaardingen, The Netherlands

P. J. Wilde, Institute of Food Research, Norwich laboratory, Norwich Research Park, Norwich NR4 7UA, United Kingdom

C. W. Wrigley, CSIRO Division of Plant Industry, North Ryde NSW 2113, Australia

PREFACE

We are living in an era of great changes. Rapid developments in communication, technology and automation open up new perspectives at a breath taking pace. In science, new technologies have emerged and are now being more and more introduced in the scientific areas "we thought we knew." The cereal chemist of today meets the molecular biologist, the spectroscopist and the statistician and learns of new possibilities for research. The cereal chemists of tomorrow will integrate these opportunities with their own skills and join in a multidisciplinary effort to solve the problems at hand.

In spite of all these changes, one thing has remained the same already for many years. It is the basic question of cereal research: "What is quality?". It is perhaps ungratifying, but nevertheless true that even today, in this spectacular era of scientific power, we have not solved the basic problems of quality control and wheat quality variation. Yes, we know a lot more than 40 years ago. An intriguing picture is emerging from the detailed information gathered by cereal scientists. We have achieved considerable progress, in some areas even dramatic. We are now able to introduce new properties into wheat by genetic manipulation. But we have not solved the questions which are most relevant to the food technologist. Food technologists today have access to all the information gathered. But can they use it? With so much sophistication and detail, it is becoming more and more difficult to preserve an up-to-date overall view. Especially difficult is the integration of information and the relation of scientific information to the day-to-day effort to produce quality.

In the present book, *Interactions, the Keys to Cereal Quality*, an effort is made to present the food technologist and cereal chemistry student of today with a more integrated view of cereal science in relation to quality. In seven chapters, the reader is taken from an overall view of cereal technology to a series of chapters providing in-depth views of the role of proteins, lipids, carbohydrates, and their interactions. Each chapter deals with a specific category of interactions taking place in wheat and its subsequent processing stages toward the final product. Each author was challenged to relate a particular field of expertise to quality. *Interaction* is the key word. In this respect, the book is quite complementary to the outstanding books *Wheat Chemistry and Technology* (vols. I and II) in which wheat constituents are described on a more individual basis. The editors feel that the focus on quality and on interactions between constituents helps readers to obtain an overall image of current-day cereal chemistry, an image which helps them understand and apply the knowledge offered by today's cereal chemists.

We wish to express our particular thanks and appreciation to mrs. Let Stevens and mrs. Huguette Bannenberg for all their work in the preparation of the book for publication

Dr. Rob J. Hamer, Prof R. Carl Hoseney

CONTENTS

CHAPTER 3

Lipid-carbohydrate interactions

CHAPTER 4

Protein-carbohydrate Interactions

CHAPTER 5

Temperature -Induced Changes Of Wheat Products

CHAPTER 1
INTERACTIONS IN WHEAT DOUGHS

W. Bushuk
Department of Food Science
University of Manitoba
Winnipeg, MB, Canada R3T 2N2

INTRODUCTION

Among the food grains, wheat is unique because it is the only grain whose flour, when mixed with water, gives a dough which can be leavened by yeast fermentation and baked into an attractive and tasty loaf on bread. Bread is a staple of human nutrition; approximately 20% of dietary calories and proteins are obtained from it. The uniqueness of wheat derives from the nature of its constituents and their ability to interact with water, themselves, and other constituents during the processing of flour into bread. This opening chapter of the book "Interactions and the Quality of Cereal Products" will be an overview of the role of interactions in the conversion of wheat into bread. The chapters that follow will address interactions within and between specific constituents of flour. The two final chapters will deal with the role of interactions in rheological properties of dough relevant to final product quality and how the interactions are modified when a bread dough is subjected to elevated temperatures during baking.

PHYSICAL AND CHEMICAL PROPERTIES OF FLOUR

Flour is made by grinding and separating the starchy endosperm of wheat (Fig. 1). During the development and maturation (desiccation) of the wheat kernel interactions among endosperm constituents occur which contribute to its texture (hardness) and thereby play a significant role in the milling of wheat into flour. These interactions are very much involved in the physical appearance of the kernel and other properties such as vitreousness or mealiness (Fig. 2). Vitreousness is an important characteristic in the grading of wheat in some countries, for example, Canada.

Hardness and vitreousness determine how the endosperm breaks up under the cutting and crushing influence of the milling rolls. The physical, and to some extent the functional, properties of flour depend on the milling process as well as on the chemistry of the flour constituents.

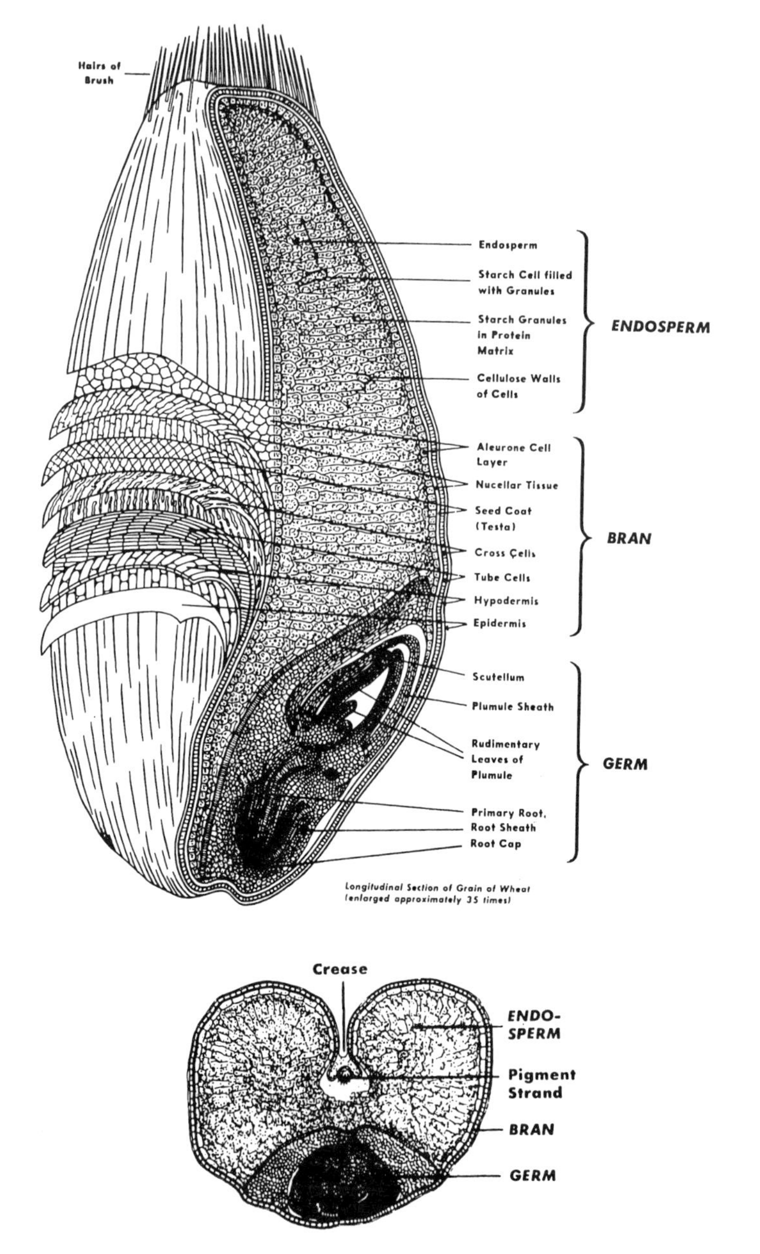

Figure 1 Schematic diagram of a wheat kernel showing the three main morphological parts. Adapted from the original by the Flour Milling Institute, Washington, DC; reprinted with permission)

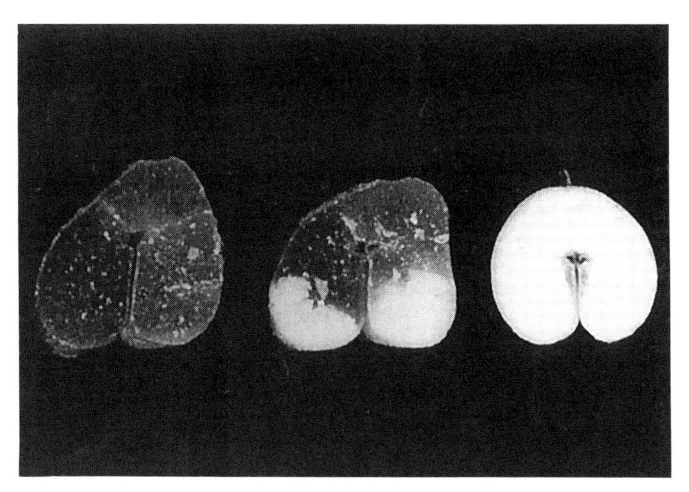

Figure 2 Cross section of wheat kernels showing varying degrees of vitreousness. Courtesy of Grain Research Laboratory, Winnipeg, MB.

Flour comprises a wide range of particle shapes and sizes (Fig. 3).

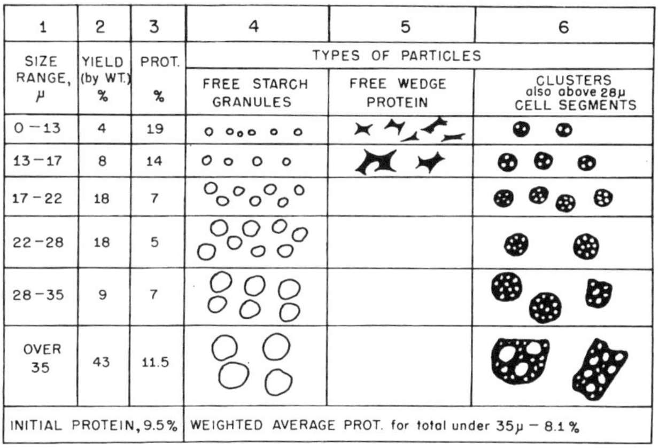

1	2	3	4	5	6
SIZE RANGE, μ	YIELD (by WT.) %	PROT. %	TYPES OF PARTICLES		
			FREE STARCH GRANULES	FREE WEDGE PROTEIN	CLUSTERS also above 28μ CELL SEGMENTS
0–13	4	19			
13–17	8	14			
17–22	18	7			
22–28	18	5			
28–35	9	7			
OVER 35	43	11.5			
INITIAL PROTEIN, 9.5%			WEIGHTED AVERAGE PROT. for total under 35μ – 8.1 %		

Figure 3 Schematic of flour particles showing particle shape, size distribution and composition. (Adapted from Jones et al (1959); reprinted with permission).

The particles vary widely in composition. Bread flour is essentially pure endosperm of bread wheat; contamination of the flour by bran and germ is usually detrimental to its use for production of white bread. Particle size (below a prescribed maximum according to the standard of identity) and size distribution determine the rate and degree of interaction on the particles with water during dough formation.

Chemical composition of flour particles varies widely from essentially pure starch or protein to that of the endosperm tissue of the wheat from which it was milled. The proximate composition of flour is shown in Table I.

TABLE I
Proximate Composition of Bread Wheat Flour

Constituent	%
Water	14.0
Starch	69.0
Protein	12.0
Lipid	2.0
Pentosan	1.8
Other	1.2

Another technologically important characteristic of flour which depends strongly on the milling process and kernel hardness is the amount of damaged starch. Flours from hard wheat contain approximately 9% damaged starch but this value can vary widely among flours (Pomeranz and Williams, 1990).

Damaged starch is important from several perspectives. It contributes significantly to the water absorption of flour; damaged starch absorbs 100% of water whereas undamaged granular starch absorbs only 30%. Also, damaged starch is highly susceptible to degradation by amylases; some of this reaction is beneficial to the breadmaking process but too much, as in flour with excessive alpha amylase activity, is detrimental and can lead to sticky doughs and even weak and sticky crumb. Water absorption of flour is important in relation to dough handling, bread yield, and post-baking bread quality (Farrand, 1964).

Specific interactions are obviously involved in the formation of the vitreous endosperm structure. A low-molecular weight (about 15 kDa) protein called friabilin is present on the surface of starch granules of soft wheats but essentially absent in starch of hard wheats (Greenwell and Schofield, 1986). After extensive studies of friabilin and its inheritance in soft wheats, it is still not certain whether this protein is the cause of endosperm softness or if it is a "marker" of other factors controlling this characteristic. For further discussion of this and relative topic, the reader should refer to chapter 5 by Marion and Clark.

Flour contains two major (starch and protein) and two minor (lipids and pentosans) components, which contribute significantly to the interactions during dough formation and development, and baking. The major components are polymers, a characteristic which is important to the viscoelastic properties of dough. The chemical composition of the interacting constituents (Table I) offers a large number and variety of interacting chemical groups that interact specifically with each other. In order for the interactions to occur, water must be added to the flour. Furthermore the mixture must be mixed (developed) to convert the flour and water into an appropriate viscoelastic dough.

INTERACTIONS WITH WATER AND ROLE OF MIXING

Flour particles are highly hygroscopic and will quickly absorb water when the two substances are brought together. Absorption of water by flour depends on the unique properties of water, e.g. its dipolar nature, and the constituents of flour. Flour constituents vary in the water uptake capacity (Bushuk, 1966). Starch (granular and damaged) take up about 46% of the water in dough because it forms a high proportion of the flour (Table II).

TABLE II
Proximate Distribution of Water in Dough

Constituent	% (d.b.)	WA (g/g)[1]	WD (%)[2]
Starch			
undamaged	56.0	0.3	18.9
damaged	24.0	1.0	27.0
Protein	14.0	2.0	31.5
Pentosan	2.0	10.0	22.5
Other	4.0	0	0

[1] Water absorption capacity
[2] Water distribution in dough

However, pentosan, a minor constituent, contributes substantially to water absorption of flour because of its high water uptake capacity; it absorbs about 10 times its own weight of water. It is well known that flours milled from different wheat varieties differ in their water absorption capacity but it is not clear whether the so-called protein "quality" factors contribute significantly to this difference.

A fascinating series of events occurs when flour particles come into contact with water. Microscopy studies, first by Bernardin and Kasarda (1973) and later by Amend and Belitz (1989), have shown that upon wetting flour particles exude visible strands or fibrils (Fig. 4). Amend and Belitz (1989) have proposed that the strands result from the explosion of hydrated protein films caused by surface tension forces. The strands are essentially pure protein with amino acid composition the same as that of gluten. Evidence has been obtained (Amend and Belitz, 1989) which showed that the strands quickly interact with each other and thereby form physical crosslinks between flour particles. It is presumed that this aggregation is the first event in the cohesion of flour particles into a dough.

Another key observation relative to interactions in dough was reported many years ago by Olcott and Mecham (1947). Specific flour lipids (polar) are bound to specific proteins of flour (glutenins) upon addition of water. Mixing of the flour and water into a dough causes binding of additional lipids (Table III). This highly specific binding of lipids is considered to be a key factor in the formation and development of a dough with appropriate rheological properties for conversion into bread.

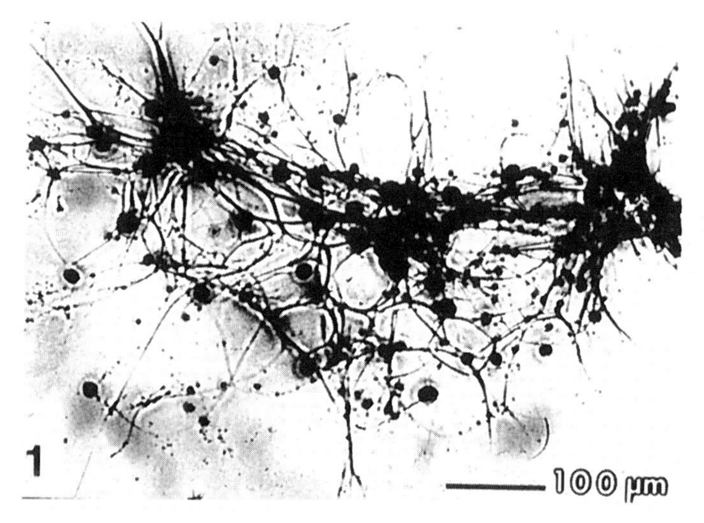

Figure 4 Protein strands which exude from flour particles on contact with water. Original figure provided by Dr. Amend, Munich Technical University.

Dough development is the most critical step in the conversion of flour into bread regardless whether it is achieved by intense mixing or by the more gentle action of extended fermentation. Improper development can destroy the baking potential of the best flour (Paredes-Lopez and Bushuk, 1983). On the other hand, appropriate development can be used effectively to correct some apparent deficiencies in the quality of the flour.

Dough mixing has several functions in the breadmaking process. The first is to blend the ingredients into a reasonably homogeneous mixture (dough). It is during the early stage of mixing that the flour particles become hydrated and gradually cohere together into an undeveloped (undermixed) dough. Mixing aids hydration by exposing new surfaces on flour particles for interaction with water. As mixing proceeds, flour particles lose their identity and the dough takes on a relatively homogeneous appearance as dough development proceeds. Dough development, reflected by a rise in dough consistency, can be followed during mixing by recording the torque on the mixer blades (as in the Farinograph or Mixograph) or the power consumed by the mixer/developer. Under appropriate mixing, i.e. above the critical speed (Kilborn and Tipples, 1972), the mixing curve will peak at a time characteristic of the cultivar indicating optimum development (Fig. 5).

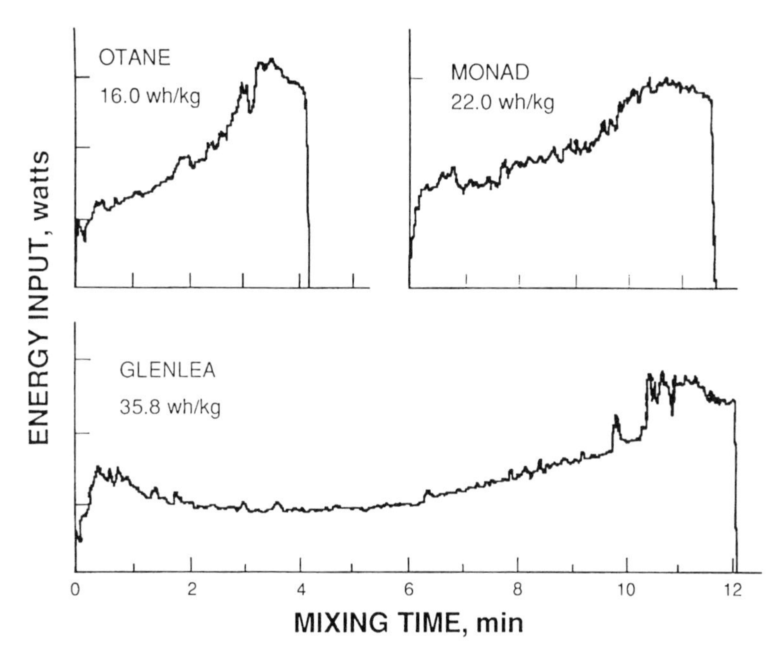

Figure 5 Mechanical development of bread doughs from wheat cultivars of widely diverse strength. Number is the amount of energy input in watt-hr per kilogram for optimum development.

TABLE III
Extractability and Distribution in Gluten Fractions of Flour Lipids[1]

	of flour %	of lipids
Alcohol-extractable fractions	1.45	100
Ether-extractable fractions (EEF)	1.0	70
EEF-flour stored at 20% H_2O contents	1.0	70
EEF-flour brought to 30% H_2O contents	0.57	39
EEF-flour wet, then dried	0.55	38
EEF-flour doughed, then dried	0.09	6

	Fraction recovered %	Lipid content %	Lipid as percent of total lipid %
Glutenin	46	20.0	81.5
Middle fraction	13	11.2	13.0
Gliadin	41	1.5	5.5

[1]Adapted from Olcott and Mecham (1947)

In terms of changes in flour constituents during dough development, there is a significant change in lipid binding (see Table III) and a marked increase in the amount of protein that is soluble in dilute acetic acid (Tanaka and Bushuk, 1973). The increase in protein solubility has been attributed to physical disaggregation of glutenin from its highly compact structure in the protein bodies in flour to a more open structure in dough (Graveland et al, 1993). Dough development is visualised as a re-orientation of glutenin polymers to form a membrane network with viscoelasticity and gas retaining properties required by bread dough. The process of protein solubilization would continue beyond peak development if mixing is not stopped, resulting in dough break-down. It is generally believed that dough development involves both physical and chemical interactions (refer next section).

Flours milled from different wheat cultivars can vary widely in the work input required for optimum dough development (Fig. 5). The fundamental reason for this is not fully understood, however it is believed that inter-protein interactions are very much involved in the observed differences between cultivars.

CHEMICAL BONDS INVOLVED IN THE INTERACTIONS IN DOUGH

Both covalent and noncovalent bonds are involved in dough structure and hence in the interactions that occur during dough formation and development.

These will be reviewed from the perspective of the flour constituents.

Starch contains two types of covalent bonds - those within the glucose monomers and the glucosidic bonds that join glucose units into the polymeric molecules of amylose and amylopectin. One of the two glucosidic bonds, the α(1-4) bond, which forms the backbone chain of the starch molecules is susceptible to hydrolysis by alpha and beta amylases. The two enzymes are naturally present in flour and therefore the bond which they hydrolyze plays an important role in the functionality of dough. The marked drop in dough consistency when alpha amylase is added to dough or as in a dough of flour milled from sprout-damaged wheat has been largely attributed to the hydrolysis of α(1-4) bonds. To what extent, if any, these bonds are broken during the milling process is not known. The glucosidic bonds in damaged starch appear to be more highly susceptible to amylase attack than the same bonds in undamaged starch.

Some of the α(1-4) bonds are continuously broken in dough during the breadmaking process to produce fermentable sugars until the amylases are inactivated during baking.

Proteins contain three types of covalent bonds:

1) those within the amino acids which normally remain unchanged during breadmaking,
2) those between amino acids (peptide bonds), and
3) disulfide bonds within and between polypeptide chains. The disulfide bonds and, in some cases, the peptide bonds play an active role in dough and bread structure.

The contribution of peptide bonds to the rheological properties of dough can be easily demonstrated by adding a protease to a dough while it is being mixed in a Farinograph. In baking practice, so-called "bucky" doughs can be mellowed by the addition of a protease preparation of bacterial or fungal origin. However, excessive indigenous protease activity as in sprout-damaged wheat can result in a drastic deterioration of the rheological properties of dough that are required for optimal baking performance.

Evidence has been published which indicates that some peptide bonds in flour proteins are more susceptible to proteolytic hydrolysis than others (Lukow and Bushuk, 1984).

The disulfide bonds of flour proteins play a key role in the interactions in doughs. They form relatively strong crosslinks within and between polypeptide chains and also stabilize other less energetic bonds such as hydrogen bonds and hydrophobic interactions. Under normal dough mixing conditions, the interpolypeptide disulfide bonds can be mobilized through the disulfide interchange reaction (Goldstein, 1957). This interchange requires a "mobile" (soluble or low molecular weight) sulfhydryl group to initiate the chain of disulfide interchanges (Fig. 6).

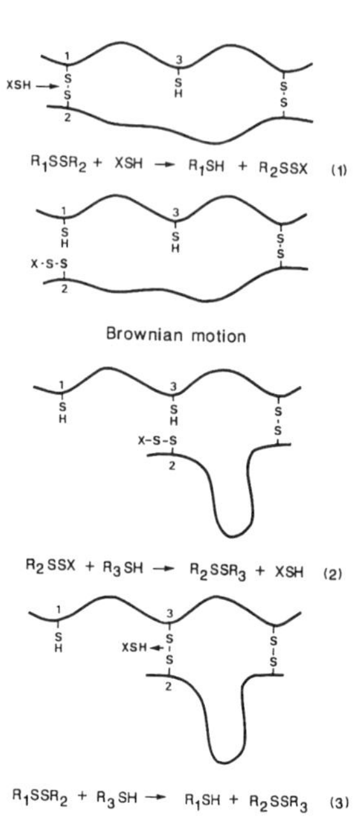

Figure 6 Hypothetical re-orientation of protein chains in dough under the influence of mixing stress through disulfide interchange reaction initiated by a mobile sulfhydryl compound. Adapted from Bloksma and Bushuk (1988).

The importance of the disulfide interchange reaction in development and stress relaxation of bread doughs cannot be overemphasised (for review see Bloksma and Bushuk, 1988).

Bakers have taken practical advantage of the disulfide interchange reaction in baking technology. By adding minute quantities of reducing substances such as cysteine, doughs can be developed much faster at lower mixing speeds. But such developed dough are extremely fragile and must be stabilized by addition of larger amounts of a slow-acting oxidizing agent such as potassium bromate. Such an oxidizing agent would be present in the dough in the early oven phase and available to inhibit further interchange reactions at a time when the dough is under considerable stress due to the oven rise with increasing temperature. In dough development, sulfhydryl groups, along with appropriately rheologically effective disulfide bonds, facilitate the alignment of protein chains by mixing to form a structure which is optimal for a particular baking process (see Fig. 6). When the reactive sulfhydryl groups are removed by oxidation as in the early oven phase, the disulfides provide the required stability for the protein matrix until the loaf structure is "set" by the gelatinization of the starch and thermal denaturation of the gluten proteins.

The next group of functionally important interactions in dough involve the so-called noncovalent bonds, hydrogen bonds, hydrophobic interactions, ionic bonds, and van der Waals bonds.

Hydrogen bonds result from electrostatic attraction between partially electropositive hydrogen atoms and the partially electronegative charge on the hydrogen acceptor. Hydrogen bonds are considerably weaker than any covalent bond but, because of the large number of bonds that can act cooperatively, they contribute significantly to the structure of dough, especially if they are stabilized by strategically located covalent crosslinks.

A unique feature of the hydrogen bond, which is perceived to be important to its functionality in dough, is its ability to interchange with other hydrogen bonds when subjected to tensile stress. This interchange would facilitate reorientation of protein chains, as in dough development, and allow for stress relaxation of structurally-activated doughs such as freshly molded doughs.

The evidence for the presence of hydrogen bonds in dough is as follows: flour proteins contain a high proportion of glutamine; starch and pentosans contain a large number of hydroxyl groups; viscoelastic doughs are formed with water not other liquids; and hydrogen bond breaking agents (urea) have a drastic effect on dough rheology.

The most compelling experimental evidence for the presence of hydrogen bonds in dough is the significant dough strengthening effect of heavy water compared with that of ordinary water as shown by the farinogram (Kretovich and Vakar, 1964; Tkachuk and Hlynka, 1968; Fig. 7). The strengthening effect results from the fact that deuterium bonds are considerably stronger than hydrogen bonds.

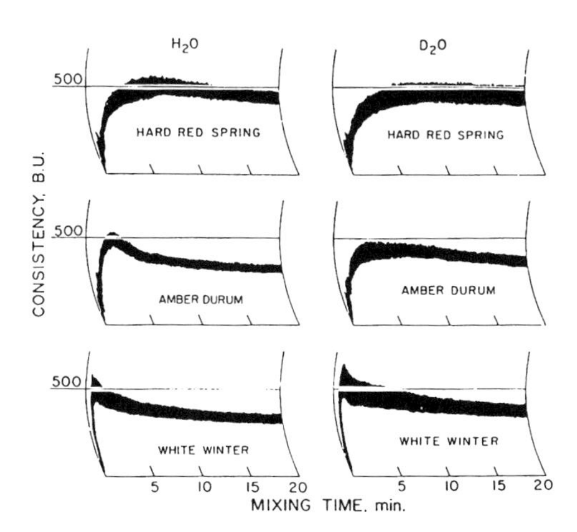

Figure 7 Farinograms of flours mixed with ordinary (H_2O) and heavy (D_2O) water indicating the presence of hydrogen bonds in dough. Reproduced with permission Tkachuk and Hlynka (1968).

The second important noncovalent bond in dough is the hydrophobic bond (interaction). This bond results from the tendency of nonpolar groups of flour constituents to associate with each other in the presence of water. The involvement of hydrophobic interactions in dough structure was originally inferred from the high content in flour of amino acids with nonpolar side chains (Bushuk, 1965). Additional evidence has been accumulated based on effects of organic solvents on rheological properties, NMR results, solubility of gluten in soap solutions, and estimates of hydrophobicity from the amino acid composition of flour proteins. When stabilized by covalent crosslinks, hydrophobic interactions can contribute significantly to the structure of dough. And, because they are freely reversible, they can readily accommodate viscous flow and thereby facilitate mechanical dough development. Hydrophobic bonds are different from other bonds in dough because their energy increases with increasing temperature; this could provide additional stability during the oven phase when the dough is subjected to high tensile stress. Another type of noncovalent bond present in dough is the ionic bond which results from the electrostatic attraction of oppositely charged groups. Ionic groups of the same charge lead to repulsion and thereby inhibit aggregation. Amino acid composition data for flour proteins indicates that the number of ionizable side chains in these proteins is relatively small. Nevertheless ionic interactions are clearly implicated in dough structure from the well-known effect of salt on rheological properties. Common salt is added to bread doughs to improve their handling properties along with taste. In terms of bond energy, ionic bonds are very close to covalent bonds and therefore a small number would contribute significantly to dough structure.

Finally a brief comment on van der Waals interactions. These interactions exist between all atoms when they are located in close proximity. These interactions produce bonds of very low energy and the energy decreases rapidly with the separation distance. Such interactions have been implicated in the structure of starch-lipid complexes but are not considered important in dough structure. Table IV gives a summary of the chemical bonds involved in the interactions in doughs.

TABLE IV
Chemical Bonds in Dough

Bond	Energy kcal.mol^{-1}	Mobility	Effect of temperature
Covalent	30-100	nil	-ve
Noncovalent			
Ionic	10-100	medium	-ve
Hydrogen	2-5	high	-ve
Hydrophobic	1-4	high	+ve
Van der Waals	0.5	high	-ve

BAKING

During baking, the temperature of the dough is gradually increased to 100°C in the interior and higher in the crust. Three important changes occur during this phase of the breadmaking process.

1. The loaf expands by approximately 50% due to the additive effects of carbon dioxide production, and evaporation of water, carbon dioxide and ethanol. This expansion involves many of the chemical bonds discussed above. The membranes surrounding the gas cells must withstand expansion without breaking in order to maintain the gas-retaining properties of the dough until its structure is set. Loaf volume depends on the ability of the dough to expand without loss of carbon dioxide. It is at this stage of the baking process at which the investment of careful development of the dough during mixing and fermentation returns the dividend of an attractive loaf of bread of good volume and texture.

2. During baking, the predominantly viscous dough is transformed into the predominantly elastic bread crumb, and at much higher temperature on the surface, into the crisp crust. Two key events contribute to this transformation are gelatinization of the starch, involving mainly noncovalent bonds, and thermal denaturation of gluten, involving both non-covalent and disulfide bonds.

3. Finally, during baking, the dough is transformed from a foam structure, in which the gas cells are self-contained, into a sponge structure in which the gas cells are interconnected. The transformation from a foam into a sponge results from the rupture of dough membranes, caused by the rapid increase in tensile stress resulting from the increase in excess internal pressure and dough viscosity. The rupture of the membranes occurs at the end of oven rise and is evidenced by a sudden increase in the loss of carbon dioxide from the loaf.

Loaf volume therefore is a direct consequence of the balance in time between oven rise and the rupture of the cell membranes; if oven rise continues for a long time a large loaf volume results.

For a more comprehensive discussion of the key chemical reactions and interactions in dough during the breadmaking process, the reader should refer to the comprehensive review by Bloksma and Bushuk (1988).

REFERENCES

AMEND. T., and BELITZ, H.-D. 1989. Microscopical studies of water/flour systems. Z. Lebensm. Unters Forsch. 189:103-106.

BERNARDIN, J.E., and KASARDA, D.D. 1973. Hydrated protein fibrils from wheat endosperm. Cereal Chem. 50:529-536.

BLOKSMA, A.H., and BUSHUK, W. 1988. Rheology and chemistry of dough. Pages 131-217 in: Wheat: Chemistry and Technology, Vol. II, Y. Pomeranz, ed. American Association of Cereal Chemists, Inc., St. Paul, MN.

BUSHUK, W. 1965. The possible role of hydrophobic bonding in the structure of gluten. Abstracts, 50th annual meeting of the American Association of Cereal Chemists, April 25-29, Kansas City, MO.

BUSHUK, W. 1966. Distribution of water in dough and bread. Baker's Dig. 40(5):38-40.

FARRAND, E.A. 1964. Flour properties in relation to the modern bread processes in the United Kingdom, with special reference to alpha-amylase and starch damage. Cereal Chem. 41:98-111.

GOLDSTEIN, S. 1957. Sulfydryl- und Disulfidgruppen der Klebereiweisse und ihre Beziehung zur Backfahigkeit der Brotmehle. Mitt. Geb. Lebensmittel Unters Hyg. 48:87-93.

GRAVELAND, A., HENDERSON, M.H., and PAQUES, M. 1993. Characterisation of glutenin proteins. Cereal Foods World 38.601.

GREENWELL, P., and SCHOFIELD, J.D. 1986. Starch granule protein associated with granule softness in wheat. Cereal Chem. 63:379-380.

JONES, C.R., HALTON, P., and STEVENS, D.J. 1959. The separation of flour into fractions of different protein by means of air classification. J. Biochem. Microbiol. Technol. Eng. 1:77-98.

KILBORN, R.H., and TIPPLES, K.H. 1972. Factors affecting mechanical dough development. I. Effect of mixing intensity and work input. Cereal Chem. 49:34-47.

KRETOVICH, V.L., and VAKAR, A.B. 1964. Effect of D_2O on the physical properties of wheat gluten. Proc. Acad. Sci. USSR, Sect. Biochem. 155:71-72.

LUKOW, O.M., and BUSHUK, W. 1984. Influence of germination on wheat quality. I. Functional (breadmaking) and biochemical properties. Cereal Chem. 61:336-339.

OLCOTT, H.S., and MECHAM, D.K. 1947. Characterization of wheat gluten. I. Protein-lipid complex formation during doughing of flours. Lipoprotein nature of the glutenin fraction. Cereal Chem 24: 407-414.

PAREDES-LOPEZ, O., and BUSHUK, W. 1983. Development and "undevelopment" of wheat dough by mixing: Physicochemical studies. Cereal Chem. 60:19-23

POMERANZ, Y., and WILLIAMS, P.C. 1990. Wheat hardness: Its genetic, structural, and biochemical background, measurement, and significance. Adv. Cereal Sci. Technol. 10:471-548.

TANAKA, K., and BUSHUK, W, 1973. Changes in flour proteins during dough mixing. I. Solubility results. Cereal Chem. 50:590-596.
TKACHUK, R., and HLYNKA, I. 1968. Some properties of dough and gluten in D_2O. Cereal Chem. 45:80-87.

CHAPTER 2
PROTEIN-PROTEIN INTERACTIONS - ESSENTIAL TO DOUGH RHEOLOGY

C. W. Wrigley, J.L. Andrews, F. Bekes, P.W. Gras,
R.B. Gupta*, F. MacRitchie and J.H. Skerritt*

CSIRO Division of Plant Industry,
Sydney NSW 2113 and
* Canberra ACT 2600,
Australia.

INTRODUCTION

Proteins are essential to many functional properties of cereal grains, and interaction between the protein components is a principal operative mechanism by which such functions occur The obvious target for studies of protein-protein interactions is gluten, the viscoelastic mass that gives wheat dough its unique properties. Without the range of interactions essential to gluten structure, it would not be possible to produce the range of leavened breads, noodles and related products that are characteristic of wheat-based foods.

In this chapter, the stronger, covalent interactions will first be described, together with evidence for the consequent importance of molecular-weight distribution to dough properties. Then follows discussion of the role of the weaker, non-covalent interactions. Elucidation of the detailed mechanisms of these many interactions is providing a basis for the postulation of models to describe gluten function in terms of specific interactions. Such knowledge should permit more predictable manipulation of product quality throughout the production chain - in breeding, in agronomic management, during storage and transport, and at the various stages of processing and marketing.

Proteins contribute to the main three quality attributes that determine the suitability of a wheat sample for processing into any of the many products made from wheat, namely, the protein content of the wheat sample, its dough properties and the grain hardness (Wrigley, 1994). The first two refer to the importance of protein quantity and quality (classically defined by Finney and Barmore, 1948), and to the main topic of this chapter. The third of these attributes (hardness) is currently hypothesized to relate to interactions (or lack of them) between proteins surrounding the starch granules and the interstitial protein matrix of the endosperm (Morrison et al, 1992; Jolly et al, 1993). This aspect of protein-protein interactions is determined primarily during grain filling, depending on genotype, with the possibility of growth and harvest conditions modifying the interactions.

COVALENT INTERACTIONS

The Formation of Covalent Bonds During Grain Filling

Also determined during the grain-filling and ripening process are, we presume, the disulfide bonding between and within polypeptides that will later be critical to the formation of gluten during dough mixing. Panozzo et al (1994a) reinforce and extend previous studies (e.g. Field et al, 1983) on the formation of storage proteins during grain development, indicating that disulfide-polymerized glutenin is extractable from grain at all stages of development, but that the synthesis of high- molecular-weight (HMW) subunits of glutenin commences a week of so before that of the low- molecular-weight (LMW) subunits. This stage of the formation of the various disulfide bonds is primary to protein-protein interactions.

Recent experiments with the expression of gluten-protein genes in tobacco (Shani et al, 1994) provide a fresh approach to studying this important stage at which presumably many aspects of protein interactions are determined. These authors concluded that, in this system at least, the assembly of glutenin subunits into insoluble polymers occurred slowly, continuing after the initiation of protein-body formation. This experimental system also provides the opportunity for testing the functional properties of novel arrangements of polymerized glutenin subunits.

Interactions Between the Major Classes of Gluten Proteins

Our current knowledge of these various disulfide bondings is the major basis for classification of the proteins of dough, as summarized in Table I.

1) The omega-gliadins, having virtually no sulfur-containing amino acids, are limited in their interactions in dough to non-covalent bonds. (Their coding genes [*Gli*-1] are on the short arms of group 1 chromosomes.)

2) The alpha-, beta- and gamma-gliadins, whose disulfide bonds are mainly formed within the polypeptide chains, are again restricted in their involvement in disulfide linkages, and thus in their ability to take part in the formation of gluten polymers. However, their SS bonds are critical in retaining the folding structure that determines the nature of non-covalent bonding. (Their coding genes [*Gli-l* and *Gli-2*] are on the short arms of group 1 and 6 chromosomes, respectively.)

3) The low-molecular-weight (LMW) subunits of glutenin, have sulphydryl groups that are directed both intrachain and towards linkage with other glutenin subunits in aggregated glutenin. (Their coding genes [*Glu-3*] lie on the short arms of group 1 chromosomes, close to the *Gli-l* loci; possibly also on group 6 chromosomes.)

4) The high-molecular-weight (HMW) subunits of glutenin, like the LMW subunits, form intrachain and interchain disulfide bonds in native glutenin,

being more effective than the LMW subunits in their contributions to dough strength. (They are also distinct from the LMW subunits in the location of their coding
genes [*Glu-l*] on the long arms of group 1 chromosomes.) The HMW subunits generally have cysteine groups available for interchain cross linking at both ends of the molecule, whereas some LMW subunits may only have this possibility at one end (the C terminal).

TABLE I
Gluten Protein Classification Based on Disulfide Bonding

Class	Locus	SS bonding
ω-gliadins	*Gli-1*	No SS bonds
α, β gliadins	*Gli-2*	}Intrachain SS bonds
γ-gliadins	*Gli-1*	
LMW subunits of glutenin	*Glu-3*	}Interchain SS bonds
HMW subunits of glutenin	*Glu-1*	
CM Proteins, etc		SS interchange?

5) The various sulfur-rich small proteins of dough, such as the CM proteins, soluble in chloroform- methanol (coded by genes on chromosomes 4 and 7) are incorporated into gluten during dough development by hydrophobic interactions and probably by disulfide interchange.

6) The general cytoplasmic proteins (albumins and globulins) of the endosperm cells may also become involved in some of the protein-protein interactions that relate to gluten function, but their role is not generally considered to be so significant as that of the other classes above.

The Importance of Disulfide Bonds

The over-riding importance of disulfide bonds to gluten function can readily be demonstrated by the addition to dough of chemical agents that break (or even just block) disulfide bonds. This is illustrated in Figure 1, which shows the drastic effects of only a tiny amount of reducing agent (0.9 mg dithiothreitol for dough made from 2g flour) on dough strength, the signs of weakening being a lower peak (note altered scale of vertical axes), a shorter time to peak resistance, and a more rapid drop in resistance to mixing after the peak (greater breakdown/less tolerance to over mixing).

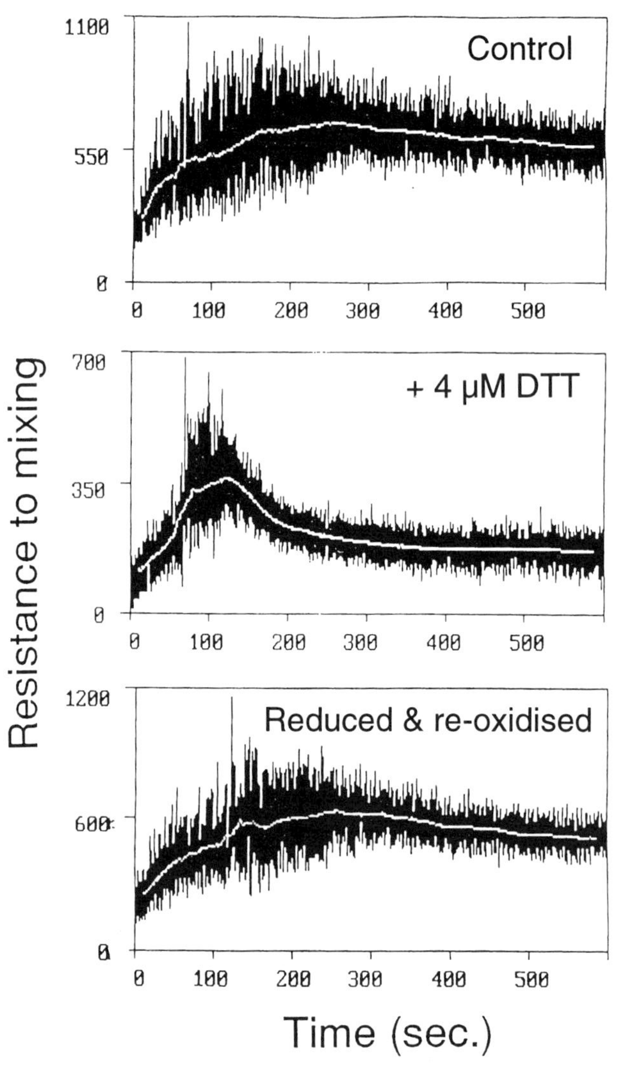

Figure 1. Addition of dithiothreitol (DTT) to dough in the two gram Mixograph and reoxidation to original consistency. To control dough (top) was added DTT (0.9 mg in 1.2 ml water, 4M) to give the second curve. The third curve was obtained by reoxidation after the DTT treatment by the addition of 0.54 micro equivalents of potassium iodate, by the procedure of Bekes et al (1994a). Note the different scaling in the middle plot for both horizontal axis (time in seconds) and the vertical axis (resistance to mixing in arbitrary units).

Further evidence of the importance of disulfide bonds to dough function is shown in Figure 1 (part three) by the addition of an oxidising agent to the partly reduced dough, producing a restoration of the original dough-strength character-istics of the mixing curve. This cycle is accompanied by a decrease in average molecular weight, followed by a return to the original molecular-weight distribution, as reported by Bekes et al (1994a).

Various studies have demonstrated that some disulfide bonds are more important rheologically (more accessible) than others (Jones et al, 1974; Moonen et al, 1985; Ewart, 1988; Ng et al, 1991). Further experimental evidence has indicated that most inter-chain junctions between glutenin subunits consist of a single disulfide bond (Kawamura et al, 1985; Ewart, 1988), leading to a general hypothesis that aggregated glutenin is a linear molecule. This linear structure may largely consist of HMW glutenin subunits, with LMW subunits branching from the polymer, also by disulfide bonds (Graveland et al, 1985; Moonen et al, 1985).

The critical importance of disulfide bonding continues beyond initial dough mixing, during proofing, and into baking, as further described in the chapter by Weegels and Hamer. During baking, the heat-induced formation of more disulfide bonds helps to set the gluten matrix as a part of the various reactions that produce the final product - leavened bread. Such is probably the mechanism by which moderate heating of dough leads to increased resistance to extension and reduced extensibility (Belitz et al, 1990) and probably also to similar changes occurring (slowly) during the long-term storage of wheat grain (Gras et al, 1994). Similar disulfide bonding has been demonstrated to accompany heat-induced damage during the commercial drying of wet gluten (Weegels et al, 1994), together with the exposure of hydrophobic groups of the glutenin fraction causing aggregation and irreversible conformational changes.

Molecular-weight Distribution and Dough Properties

The weakening of dough (Fig. 1) associated with disulfide-bond rupture is accompanied by a general decrease in the molecular-weight (MW) distribution.

This is demonstrated by complete extraction of protein from dough (using the combination of sonication and sodium dodecyl sulfate (SDS) to break non-covalent bonds, and yet minimise the rupture of covalent bonds) and fractionation of the extracted proteins on the basis of MW by size-exclusion high performance liquid chromatography (SE-HPLC) (Fig. 2).

Figure 2 also shows the effect of bromate addition in increasing protein-protein interactions to the extent of extending size distribution with a small peak in the highest size range (extreme left of profile). This may be (at least part of) the mechanism of baking quality improvement for which bromate has long served (Panozzo et al, 1994b).

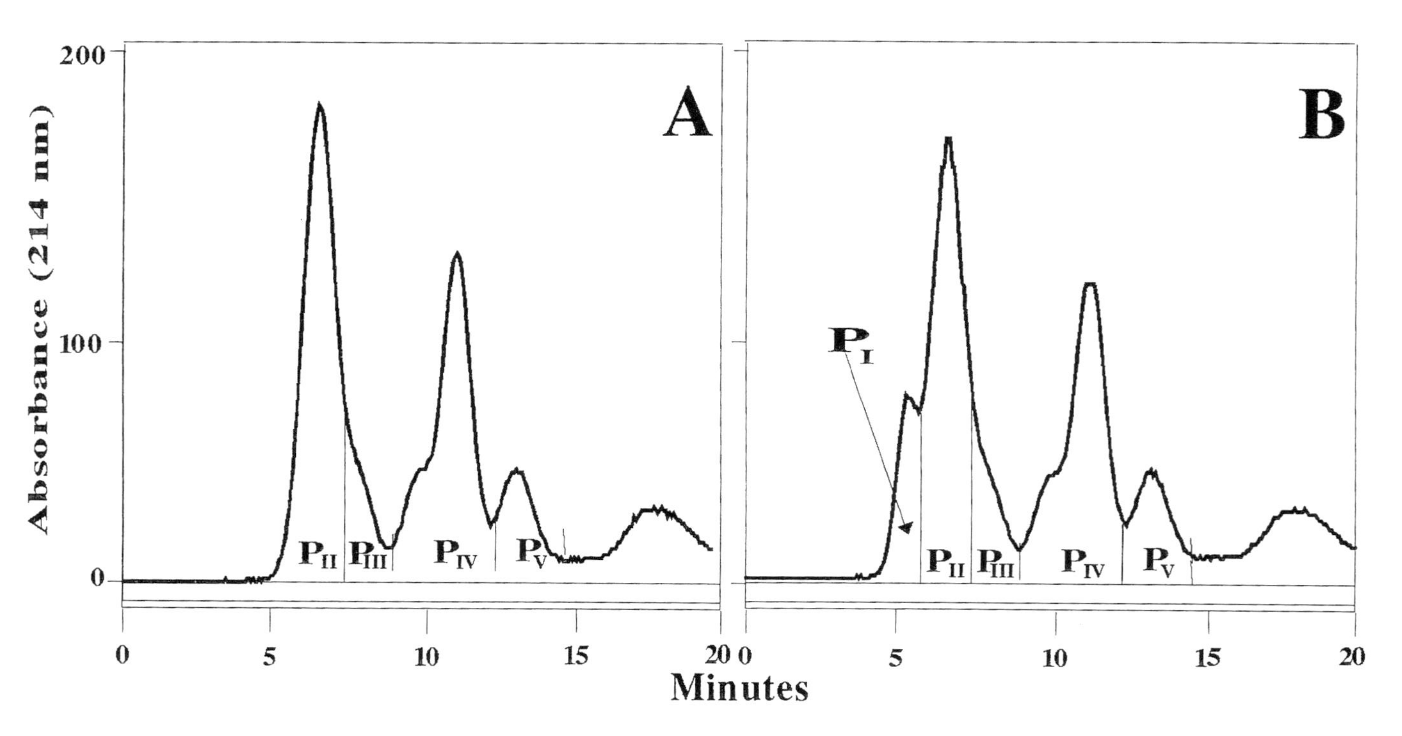

Figure 2. SE-HPLC of dough protein extracted with sonication into SDS-containing buffer to indicate MW distribution, without (left) and with (right) the addition of 10 ppm of bromate, according to the procedure of Panozzo et al (1994b). P_I and P_{II}=glutenin, P_{III}=gliadin, P_{IV} and P_V=soluble proteins. Adapted from Panozzo et al (1994b).

On the other hand, it has long been known that during the mixing of dough, there is an increase in the general solubility of the dough proteins, due to the breaking down of the large macro molecular structure (Tanaka and Bushuk, 1973). Graveland et al (1980) demonstrated this in relation to the solubility of gluten proteins in solutions of the detergents sodium dodecyl sulfate (SDS), going so far as to postulate the involvement of the superoxide anion in the reduction of SS bonds to sulfhydryl groups.

During the past two decades, there have been attempts to quantitate the very large polymers of glutenin by solubility (e.g. residue protein of O'Brien and Orth 1977), in the absence of techniques to adequately analyse for proteins with molecular weights into the millions. Linked to the quantitation of the less soluble fraction of gluten is the concept of gel protein - that part of the gluten protein that separates as a gel layer during centrifugation of a flour-SDS extraction mixture (Graveland et al, 1982; Pritchard and Brock, 1994). Findings that the amount of gel protein relates to dough quality have led to its use in various forms as rapid tests for quality. Underlining both the concepts of insoluble residue and gel protein is the existence of very large covalently linked polymers in the gluten complex.

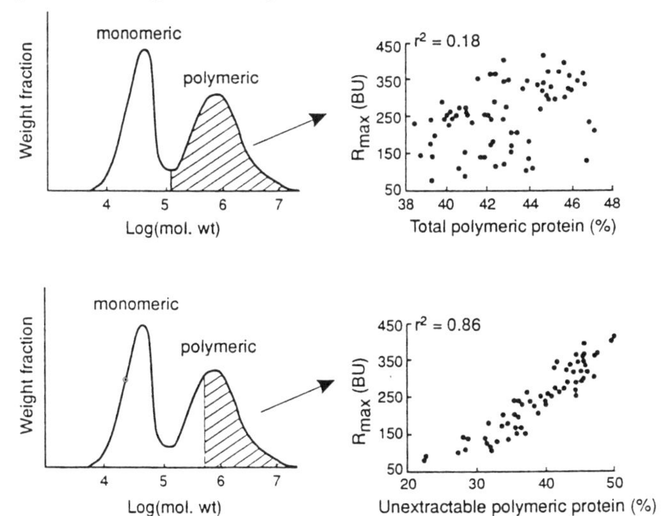

Figure 3. Relationships between Extensigram height (R_{max}) and glutenin fraction, either total polymeric protein (% of protein for a range of recombinant inbred wheat lines) (upper pair of illustrations) or the larger size range of this fraction (as the % of protein only extractable into SDS-buffer after sonication). Adapted from Gupta et al (1993).

In a more recent attempt to probe the upper limits of this size range, Gupta et al (1993) devised a double-extraction procedure, using SDS-containing buffers (without reducing agent), first without sonication and secondly with this physical means of releasing the larger polymers for subsequent quantitation by SE-HPLC Figure 3 shows an example of the resulting improvement in relationship to R_{max} by focusing on the more difficult-to-extract polymers of glutenin (r^2 changing from 0.18 to 0.86 for the set of wheats examined). The contrast in fractions thus presumed to be examined is illustrated on the left of Figure 3 (MW increasing left to right). Such results thus indicate that attributes related to dough strength increase with a higher proportion of very large glutenin polymers.

Multi-layer Gel Electrophoresis

In an attempt to extend the direct analysis of MW distribution into the much larger size range, we have adapted an electrophoresis procedure using several layers of polyacrylamide gel (starting with a very large pore gel) (Khan and Huckle, 1992; Wrigley et al, 1993). The result provides a quantitative estimate of interactions leading to glutenin polymers in five size ranges extending up into the millions of MW. Unfortunately the lack of very large standard proteins of known size precludes the possibility of calibration of the larger size ranges.

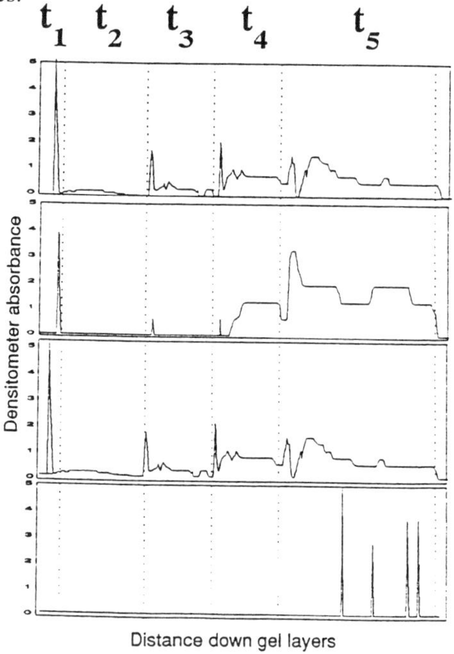

Figure 4. Multi-layer gel analysis of glutenin polymer size distribution with SS bond rupture and reformation. The top three profiles correspond to extracts of the doughs, respectively, in Figure 1. The bottom profile is for a fully reduced extract of the original dough. Adapted, from Wrigley et al (1993).

The examples of these profiles shown in Figure 4 illustrate the decreases in the proportions of protein in the largest size ranges due to partial disulfide-bond rupture and the retum to the original size distribution after the reformation of these bonds with oxidation (Bekes et al, 1994a)

Figure 4 should be studied together with Figure 1 which shows the respective weakening of dough properties accompanying the partial reduction of disulfide bond interaction and the restoration of dough properties after re-forming of the bonds. Figure 4 also illustrates the loss of all glutenin polymer structure (and of all gluten function) with complete rupture of disulfide bonds (Figure 4, bottom profile).

Multi-layer gel electrophoresis has been used to screen for MW distribution in extensive ranges of wheat-flour samples of known dough properties. For a set of 40 samples of varieties from around the world, the best correlation was provided to dough strength (as R_{max}, Extensigram height) by a weighted combination of the proportions of glutenin polymer in the five size ranges, mainly accen-tuating the largest size fraction. This is illustrated for one of the sets of samples in Figure 5. The relationship (r^2) to R_{max} was 0.71 using the largest size only, or 0.79 using a weighted estimate of MW distribution from multi-layer gel electrophoresis.

Importance of Other Covalent Interactions

A further illustration of the over-riding importance of size distribution to dough properties is provided by the weakening effect of breaking peptide bonds, either by the addition of protease to dough (demonstrated by the commercial practice of softening excessively tough doughs) and naturally for dough from bug-damaged wheat. In the latter case, the saliva of the insect *Nysius huttoni* contains a protease that causes specific hydrolysis of the peptide bonds of hIW glutenin subunits (Swallow and Every, 1991), thus weakening dough properties in a way similar to the reduction of glutenin chain length by disulfide-bond rupture. The action of the endogenous proteases of germinating grain can also cause dough weakening, as was illustrated by analyses of doughs made from germinated wheat (Bushuk and Kawka, 1990).

On the other hand, a dough-strengthening effect has been demonstrated by the synthetic cross linking of gluten-protein chains using glutaraldehyde (Wrigley et al, 1972; Simmonds et al, 1972). We demonstrated that flour giving a weak, extensible dough could be transformed into a flour with stronger dough properties by treating the flour with 40 p.p.m. of glutaraldehyde. Treatment may involve mixing the dry flour with the glutaraldehyde dissolved in solvent (which is allowed to evaporate), by adding glutaraldehyde to the water used to make the dough, or by misting the glutaraldehyde into the dry flour. The chemistry of this treatment had previously been studied by Ewart (1968), who concluded that glutaraldehyde formed cross links between free amino groups of the gluten proteins. We found that a mono-aldehyde such as butyraldehyde did not strengthen dough properties in this way, presumably because it could not form cross links between polypeptide chains.

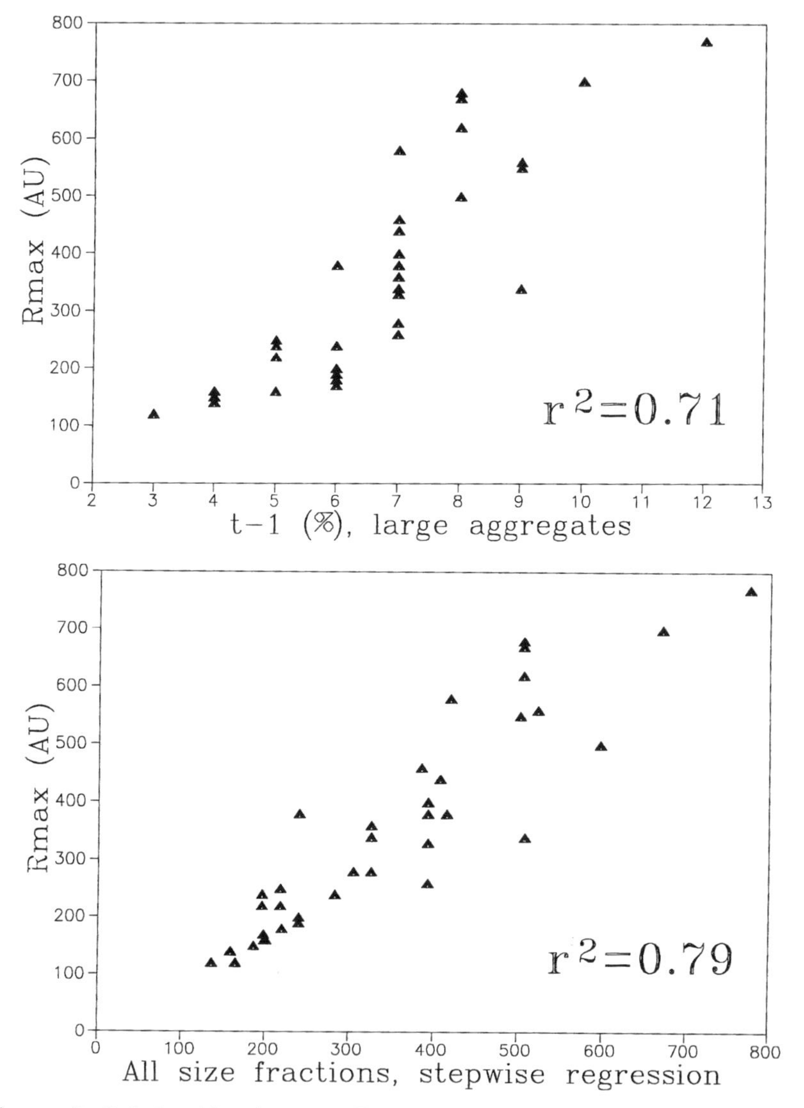

Figure 5. Relationships between Extensigram height (R_{max}) and either the proportion of extracted protein (t_1) on the top of a multi-layer gel (method of Wrigley et al, (1993) (upper graph) or a weighted average proportion based on all size fractions (t_1 to t_5) (lower graph).

NON-COVALENT INTERACTIONS

The Diversity of Non-covalent Bonds

Methods of analysis of molecular-weight distribution require firstly that non-covalent bonds must be broken to permit the extraction of the dough proteins into dilute solution/suspension for size analysis. Superficially, the size distribution of the SS-bonded glutenin polymers might appear to be the predominant determinant of dough strength, but it must be remembered that the analytical methods used have already destroyed any information about non-covalent bonds. The range and nature of these bonds is described in Chapter 1 by Bushuk and in other chapters concerning interactions of proteins with other components of dough. Such interactions involve ionic, hydrogen, hydrophobic and Van der Waals bonds, as well as lipid-protein and carbohydrate-protein interactions.

Despite the diversity in the range of types of non-covalent bonds, there were relatively small changes in dough-mixing properties caused by the addition of reagents designed to alter various non- covalent interactions. There were slight weakening effects on Mixogram traces by the addition of urea (presumably disrupting hydrogen-bonding), or of ethanol (disrupting hydrophobic bonds). These changes were minor compared to the dramatic weakening of dough by disulfide bond rupture using a much smaller concentration of DTT reagent (Fig.6). Even when the reagents presumed to affect non-covalent bonds were added at much greater concentrations (Table II), the changes in dough parameters were modest, compared to the addition of DTT.

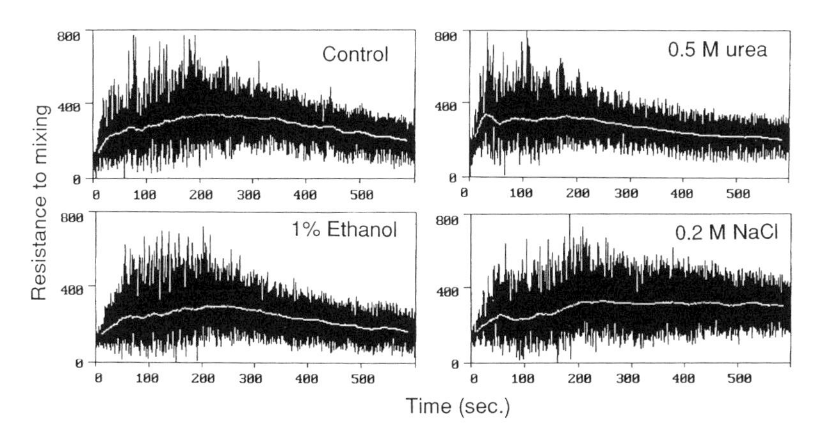

Figure 6. Mixogram traces, showing control (upper left) and additions of urea (0.5M in water added to flour; upper right), of ethanol (1% v/v; lower left) and of sodium chloride (0.2M; lower right). See Fig. 1 for identification of axes.

TABLE II
Changes in Mixo-graph determined dough properties due to the addition of reagents (as concentration in water added to 2g flour) presumed to affect non-covalent interactions

Treatment	Time to peak (sec)	Peak resistance
Control	220	366
1% ethanol	237	306
5% ethanol	142	422
10% ethanol	105	430
30% ethanol	52	407
0.1M urea	300	468
0.5M urea	193	490
1.0M urea	108	425
0.2M NaCl	240	323
0.5M NaCl	448	493
2.0M NaCl	567	467

Belitz et al (1990) investigated the role of ionic interactions by studying the effects of sodium chloride additions to dough, progressively increasing the additions from 0.5% to 10%. The strengthening effects observed at lower levels (<5%, depending on variety) were presumably due to the salting-out effect of the addition leading to greater protein-protein interaction. The stickiness and weakness of doughs at higher salt concentrations could be interpreted as disruption of ionic bonding. Such an explanation may be excessively simplistic, however, given the range of reactions to a wider range of different ionic species, as reviewed by MacRitchie (1992).

Hoseney and Brown (1983) performed mixing studies at low pH and concluded that the decrease in stability to overmixing was due to conformational changes in the gluten proteins, presumably leading to alterations in non-covalent bondings. Mixing doughs under nitrogen on the other hand, leads to increased tolerance to overmixing (Baker and Mise, 1937), possibly involving mechanisms opposite in nature to the effects of bromate.

Another approach to demonstrating a particular class of non-covalent interaction has been the use of deuterated water, shown to strengthen dough, presumably by increasing hydrogen bonding (Kretovitch and Vakar, 1964; Hoseney, 1986; Bushuk and Kawka, 1990). A more sophisticated use of deuterated water has involved NMR analyses of the hydration of glutenin samples in comparison with the mammalian protein elastin, demonstrating distinct differences between the two types of proteins (Belton et al, 1994). Recent molecular-modelling studies of Kasarda et al (1994) indicate that hydrogen bonding between glutamine side chains may be very important in stabilising the conformation of gluten proteins, particularly for the sequence-repeat domains of the HMW-glutenin polypeptides.

Gluten - One Giant Molecule?

When the effects of the full range of interactions are taken into account, the mass of gluten protein in dough might be considered to be one large macro-molecule, with all components interacting covalently and non-covalently. New methodologies are needed for significant study of the component polypeptides and amino-acid sequences involved, particularly for the non-covalent interactions. The various models that have been proposed for the structure of the gluten complex have concentrated on its covalent structure and the likely positioning of disulfide bonds (e.g. Khan and Bushuk, 1979; Graveland et al, 1985; Ewart, 1979, 1988; Kasarda, 1989; Ng et al, 1991; Gao et al, 1992). As a result, there has been less accent on the role of non-covalent interactions in such models. Although the manner of formation of covalent bonds between polypeptide chains is critical, the non-covalent interactions will probably prove to be important in explaining why some gluten polypeptides may be more effective than others of similar size in contributing to dough properties.

IDENTIFICATION OF THE POLYPEPTIDES INVOLVED IN INTERACTIONS

Electrophoretic Analyses of Interacting Components

In an earlier model of gluten structure, the importance attached to non-covalent interactions led to the suggestion that all disulfide bonding in gluten is intra-molecular, but that these bonds are critical to holding conformations that maximised non-covalent interactions; rupture of disulfide bonds thus led to loss of non-covalent bonding and to loss of rheological properties (Kasarda et al, 1978).

Evidence against this model comes partly from gel-electrophoretic analysis of unreduced gluten, extracted and fractionated in SDS-containing buffer which should disrupt non-covalent bonds. A continuous streak of glutenin polymers is obtained; rupture of disulphide bonds releases clearly resolved zones of glutenin subunits that are superimposed on the zones of proteins such as the gliadins that are not involved in the glutenin macro polymer (Fig. 7).

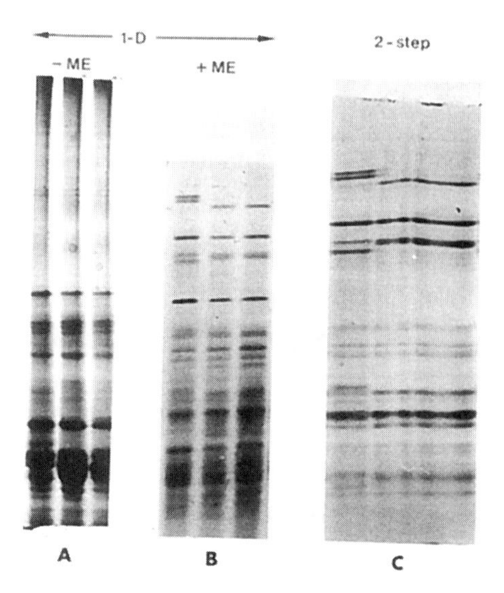

Figure 7. SDS gel electrophoresis of proteins (A) extracted from wheat flour with SDS-buffer (no reducing agent), (B) extracted with reducing agent (1.5% mercaptoethanol, ME), and (C) cut from the top of a gel such as in (A), and incubated with 1.5% ME before electrophoresis according to the procedure of Gupta and Shepherd (1990).

Attempts to correlate aspects of gluten composition to function have concentrated on these discrete zones (especially the glutenin subunits) in preference to the more difficult task of analysis of the glutenin "streak". These studies at the polypeptide level have led to considerable success in ranking the apparent contributions to dough properties of individual polypeptides, initially the HMW subunits (Payne, 1987; MacRitchie et al, 1990), and later the LMW subunits (Gupta et al, 1991; Lew et al, 1992) and the gliadins (e.g. Metakovsky et al, 1993; Wieser et al. 1994).

Study of near-isogenic and multi-null lines has provided considerable evidence about the roles and functions of the individual gluten polypeptides. Figure 8, for example, contrasts the pairs of biotypes isolated from three Australian cultivars which were natural mixtures of genotypes with either 2+12 or 5+10 HMW glutenin subunits. The results firstly reinforce the conclusion that subunits 5 and 10 are associated with greater dough strength (in this case, shown as longer time to peak development in the Mixograph). The corresponding contrast in protein composition does not relate to the total proportion of polymeric glutenin (lower right of Fig. 8), but to the amount of the very large polymeric glutenin, probably thus to the better ability of the 5 and 10 HMW subunits to polymerise into very large macromolecules (MacRitchie, 1992).

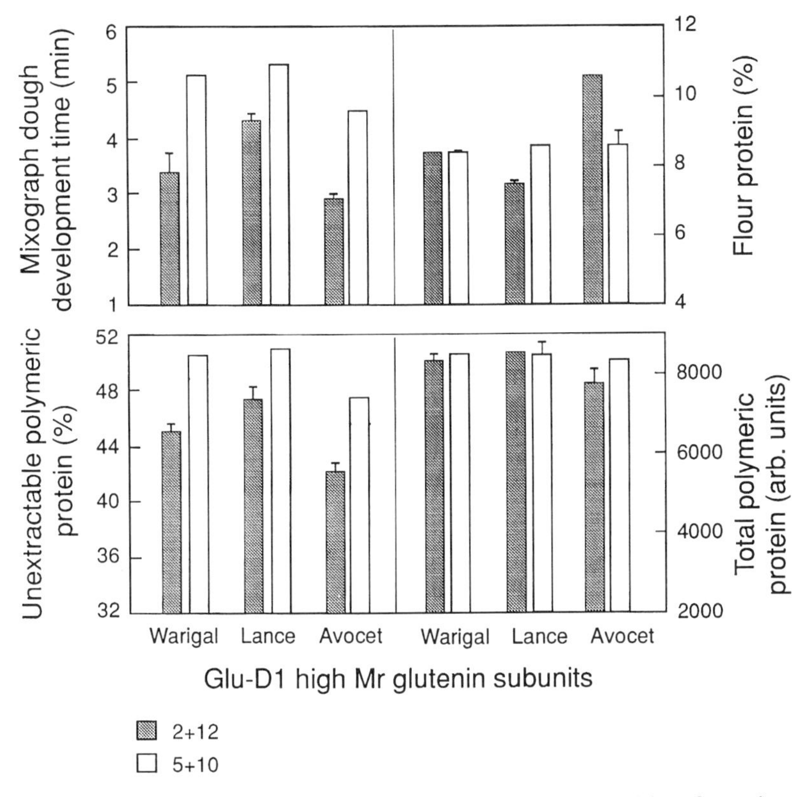

Figure 8. Contrasting dough properties and protein composition for pairs of biotypes from three Australian cultivars, in each pair one with 2+12 HMW glutenin subunits, the other with 5+10 subunits. Adapted from Gupta and MacRitchie (1994).

Although greatest research attention has focused on the HMW subunits of glutenin, the LMW subunits also play an important role in determining dough properties (e.g. Pogna et al, 1988; Gupta and MacRitchie, 1994), although the roles may differ. For example, Figure 9 illustrates that increasing the proportion of HMW subunits (in a set of multi-null lines) leads to longer mixing time, whereas a higher proportion of LMW subunits gives a shorter mixing time for a similar R_{max} (MacRitchie, 1992).

The difficult tasks now facing gluten chemists are 1) to determine whether these polypeptide- quality relationships are cause-and-effect, and 2) if they are, to elucidate the reasons for the associations. These investigations will go beyond determining the primary interactions between polypeptides to studies of their three-dimensional conformational and sequence specificities.

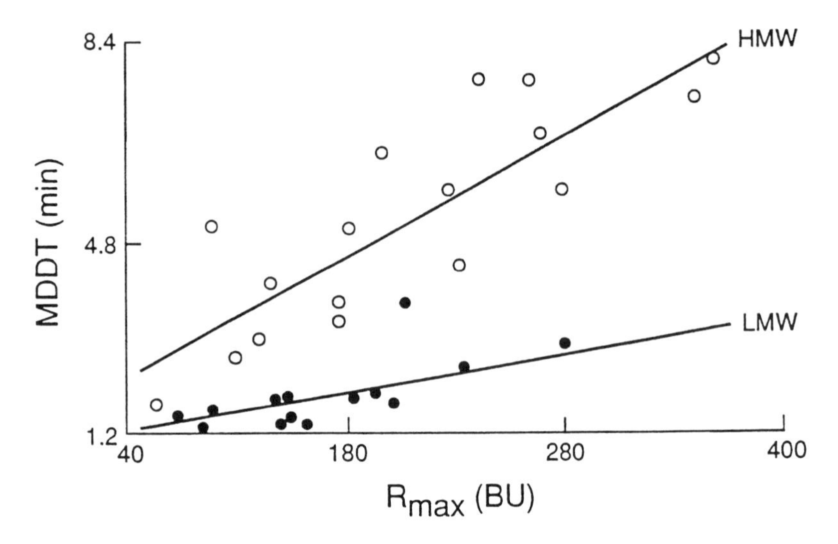

Figure 9. Relationships between Extensigram height (R_{max}) and dough development time in the Mixograph (MDDT) for sets of wheats varying in the proportions of HMW subunits (upper line, open circles) or in LMW subunit proportions (lower line, closed symbols). Adapted from MacRitchie (1992).

Testing Gluten-polypeptide Function in Dough

To address the first of these questions (are the correlations due to cause-and-effect?), procedures have been devised for incorporating isolated polypeptides into the dough-protein matrix (Bekes and Gras, 1992). These experiments have first of all required the development of very-small-scale equipment for analysis of dough properties, namely the direct-drive Mixograph. Its commercially available version is illustrated in Figure 10.

This instrument needs only two grams of flour, and thus only a few milligrams of purified protein (Rath et al, 1990).

The simple addition of protein fractions to dough caused a weakening effect, even when HMW glutenin subunits were used. Actual incorporation (or opportunity for incorporation) of the added protein into the gluten matrix was required for appropriate testing, using the cycle of partial reduction and re-oxidation illustrated in Figure 1. In this way, significant dough strengthening effects were obtained for several HMW subunits of glutenin (Figure 11), as indicated by increasing dough mixing times. This was not observed for purified gliadin proteins (Bekes et al, 1994b, 1994c).

Figure 10. The 2-gram direct-drive Mixograph (as manufactured by TMCO/ National, Lincoln NE, USA:reprinted with permission).

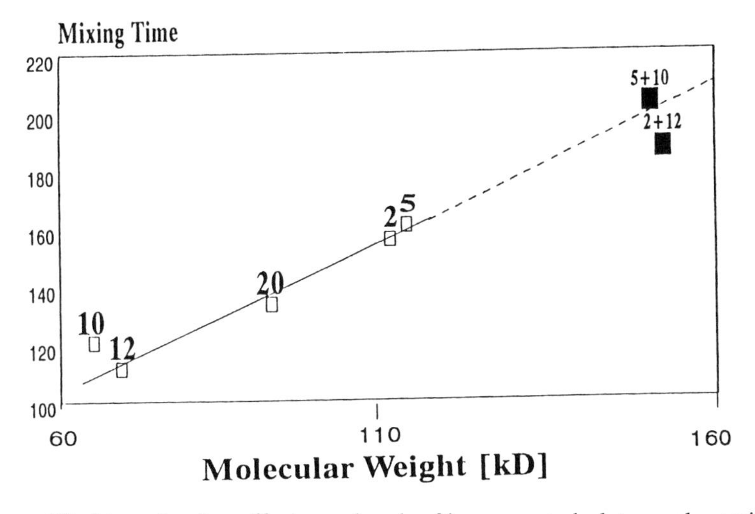

Figure 11. Strengthening effects on dough of incorporated gluten polypeptides in relation to their MW.

The extent to which the various HMW subunits affected dough properties depended particularly on their chain lengths (MW), with the combination of subunits 5 with 10 (or 2 with 12) acting as if they had been incorporated as the dipeptide in each case (Fig. 11). This suggests that these pairs may readily combine, presumably by disulfide bonding, in preference to the other interchain interactions.

The report of Tao et al (1992) has particular relevance to this result. They examined peptide fragments released by partial proteolytic digestion of insoluble glutenin aggregates by Lys-C, and found dipeptide fragments consisting of one x-type joined to one y-type subunit. Furthermore, the x- type C-terminal region was linked to a y- type N-terminal region. This corroborates the much earlier report of Lawrence and Payne (1983) of dipeptide fragments being released by incomplete reduction of glutenin. More recently, two-dimensional gel electrophoresis has also indicated the release of specific dipeptides during partial reduction of glutenin (Werner and Kasarda, 1991). Similar studies using chromatographic methods on glutenin hydrolysates (Kohler et al, 1991, 1993) have further identified the polypeptide chains that are SS cross-linked. These included inter- and intra-molecular bonding for HMW subunits. Cross linking also involved LMW subunits of glutenin and even gamma- gliadins. There is thus an emerging picture of SS interactions of gluten polypeptides not being merely random (see Figure 12).

Hybrid Polymers of Glutenin With Soluble Protein

To address the question of the uniqueness of gluten proteins with respect to their efficacy for interaction, we repeated the above addition/incorporation experiments in the two-gram Mixograph using bovine serum albumin (BSA), a polypeptide similar in size to some of the gluten polypeptides (monomer MW 67,000) having sulphydryl groupings for potential cross linking to the gluten matrix, but being water soluble and much more hydrophilic than the polypeptides of gluten.

Figure 13 and Table III show that the addition of 2 mg BSA to dough (from 2 g flour) had only a minor effect on dough properties. Incorporation (via the reduction/oxidation cycle) of 2 mg BSA, on the other hand, produced a dramatic increase in the rate of breakdown after the peak. This poor stability to over-mixing may be due to the glutenin hybrid chain becoming more soluble, with poorer non-covalent interactions between polymer chains.

It contrasts with the dough properties obtained after the incorporation of HMW subunits, when all aspects of the Mixogram trace show increased strength.

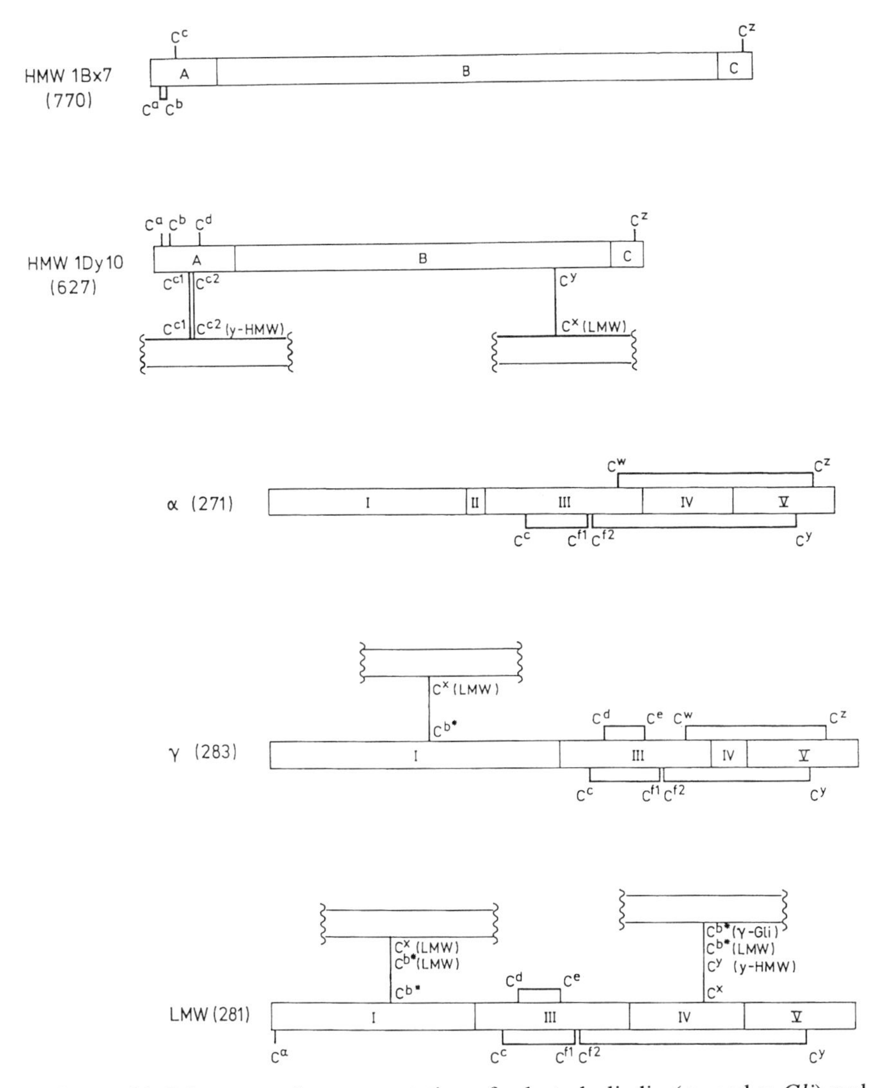

Figure 12. Diagrammatic representation of selected gliadin (α- and γ-*Gli*) and glutenin polypeptides (HMW and LMW), showing specific domains (I, II,... or A, B, ..., from Kasarda et al, 1984) and positions of intra- and inter-chain disulfide bonds at specific cysteine residues (indicated by C with a superscript). This figure is a summary, kindly provided by Drs. Wieser and Kohler, of results from Kohler et al, (1991, 1993), Keck et al (1995) and from Muller and Wieser (1995).

TABLE III

Changes in Mixogram Attributes Caused by the Addition or Incorporation of Bovine Serum Albumin (BSA) Into Dough by the Techniques of Bekes et al (1994).

Treatment	Time to peak (sec)	Peak resistance	Resistance Breakdown
Control	220	366	16
Addition of			
2 mg BSA	285	305	18
5 mg BSA	210	330	16
Incorporation of			
2mg of BSA	210	315	15

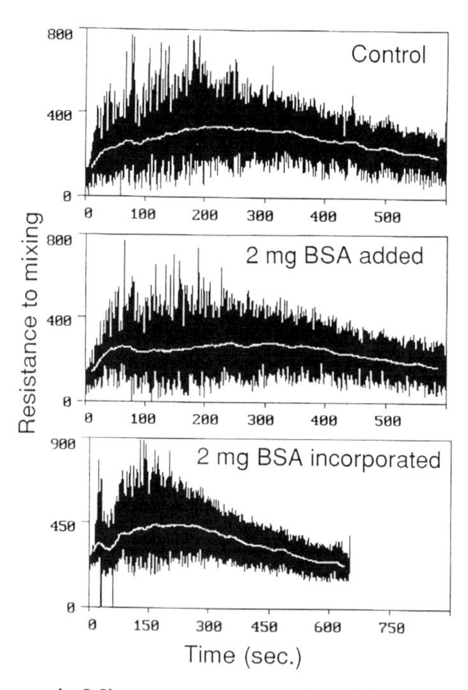

Figure 13. Changes in Mixogram traces produced by the addition (middle) of 2 mg bovine serum albumin (BSA) to the control (top) or the incorporation of 2 mg BSA (bottom trace).

DEVELOPING A MODEL TO DESCRIBE INTERACTIONS

Probing for Amino-acid Sequences Conducive to Desirable Dough Properties.

The development of antibodies whose reaction with flour proteins correlates with dough strength provides a tool for studying the amino-acid sequences that may be critical to dough-forming properties (Andrews and Skerritt 1994). These studies have indicated that polypeptide regions of high Mr glutenin subunits recognised by dough strength-related antibodies have a high probability of forming overlapping beta-turns. This agrees with structural models of glutenin subunits forming beta-spirals (Field et al, 1987; Miles et al, 1991). The consistency of beta-turns may account for some of the elasticity of dough either directly, although not in the same manner as elastin (Belton et al, 1994), or through improved interactions with other gluten components.

Entanglement Coupling of Gluten Polymers

The picture emerging from the experimental evidence described above is that dough strength increases with:

a) increasing protein content of flour;
b) increased proportions of very large glutenin polymers;
c) increased proportions of HMW glutenin subunits, compared to LMW subunits;
d) the presence of specific subunits of glutenin.

These observations are depicted in the model of interactions portrayed in Figure 14. The diagrams are based on the general property of all polymer systems, namely, that above a critical molecular size, physical properties such as tensile strength show a much steeper rate of increase with increasing molecular size. This is believed to be due to frictional resistances at widely spaced points in the polymer chain, referred to as entanglements. An entanglement network appears to be a good model for the continuous protein structure in a developed dough (Fig. 14). The entanglements act as transient cross-links, i.e. they provide a resistance when dough is subjected to a stress but, unlike true covalent cross-links, are potentially able to slip free as the stress is maintained. Non- covalent bonds can be presumed to modify and enhance the more conventional concept of entanglement coupling. They would slip to the extent that mechanical extension would cause their rupture, allowing re-formation in another conformation with chains that lie nearby.

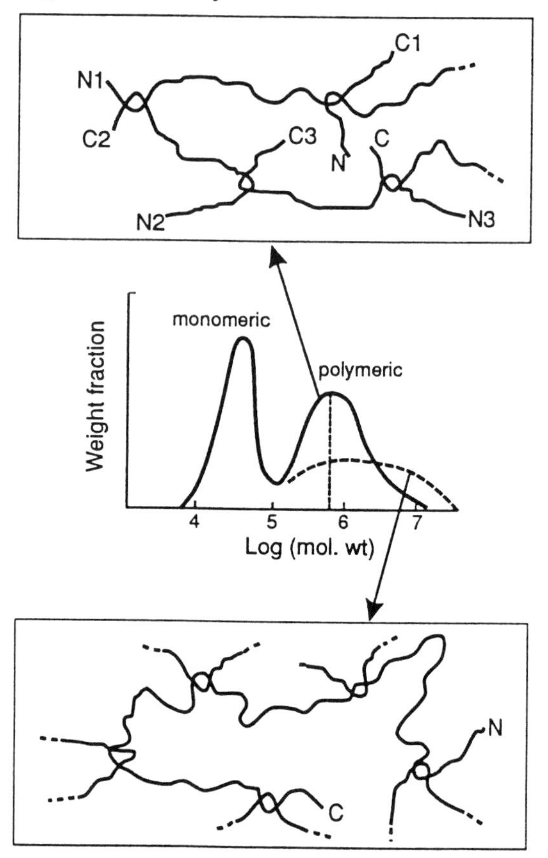

Figure 14. Model of interactions between the chains of gluten polymers due to entanglement coupling (darkened region where different chains cross). The bottom diagram (corresponding to the dotted MW profile) shows only one polymer chain (from N to C terminal ends), and entanglements with several other chains. The top diagram illustrates three short chains (N1...C1, N2...C2, N3...C.) whose entanglements with one another would be readily disentangled by applying a stress; their lower size distribution is shown by the solid line in the graph.

If the glutenin polymers are of moderate length (upper part of Fig. 14), their entanglements with one another can be readily disentangled by slippage, e.g. the C terminal end of chain 2 from the N terminal of chain 1 on the left of the upper diagram. This situation corresponds to the solid line of the MW distribution profile. By contrast, the broken line in the graph represents a gluten sample with much longer glutenin chains, one of which appears in the lower diagram. When a stress is applied in this case, disentanglement is much more difficult; therefore resistance to extension is greater, mixing time (to peak resistance) and intensity would also be greater, but extensibility would be lower. Therefore dough properties might be modified primarily by factors that alter chain length for glutenin polymers. In addition, non-covalent interactions between the chains can be seen to determine slippage of these chains against one another, thus also affecting dough properties.

IMPLICATIONS FOR IMPROVING WHEAT-GRAIN QUALITY

Better knowledge of how protein-protein interactions relate to dough properties should in turn provide a basis for more intelligent manipulation of quality attributes in breeding and in managing the growth and harvesting of the grain. For example, sulfur deficiency in the presence of adequate nitrogen fertiliser has been shown to modify dough properties (increased R_{max} and lower extensibility) by decreasing the proportions of the sulfur-rich gluten proteins (particularly the high- molecular-weight subunits of glutenin), whilst there is a higher proportion of the low-sulfur polypeptides such as the omega-gliadins (Randall and Wrigley, 1986; MacRitchie and Gupta, 1993; Schnug et al, 1993). As a result, there is much less opportunity for normal protein-protein interactions, due to a higher proportion of HMW subunits and of large polymeric glutenin. This example indicates that it is possible to have a shift in the MW distribution to a level that is too high to provide satisfactory extensibility.

Episodes of heat stress during grain filling (a few days of daily maxima over 35 °C have been shown to weaken dough properties for the resulting grain of heat-susceptible wheats (reviewed by Blumenthal et al, 1993). This change in dough properties was associated with an increase in the proportion of gliadin, thus diluting the polymeric glutenin fraction, and in yet another way altering the normal course of protein-protein interaction.

Weakness and dough stickiness have been a problem for many wheat genotypes substituted with the short arm of chromosome 1 (lRS) from rye (e.g. Fenn et al, 1994). As a result, there is a loss of some of the LMW subunits of glutenin (those coded by the wheat chromosome arm that has been substituted) and an increase in monomeric protein due to the Secalins from the lRS arm. Overall, the balance of polymeric versus monomeric protein is shifted in the direction of lower molecular-weight distribution and a decrease in dough-strength attributes. On the basis of our current understanding of composition-function relationships, strategies for redressing this balance have been recommended (MacRitchie, 1992), and successes are being achieved in doing so (e.g. Shepherd et al, 1991).

In addition to the practical implications of these examples of quality variations, they provide valuable opportunities for studying the effects on processing quality of major changes in protein composition using the whole flour, as distinct from study of flour samples reconstituted by re-combining fractions.

A completely new opportunity for studying the relationships of interaction/composition of the proteins on dough quality is being offered by the construction of modified genes for gluten polypeptides, such as those reported by Shani et al (1992). Expression of such novel genes in a heterologous system provides the opportunity to test the functional properties in dough (Bekes et al, 1994 a,b,c) for subunits having very different structures. For example, glutenin polypeptides are becoming available with a longer or shorter repetitive region in the middle of the polypeptide chain, or with different numbers and spacings of cysteine residues in the polypeptide. These approaches will contribute significantly towards formulating models of the ideal structure for the range of gluten proteins. Recent advances in the transformation of wheat (Vasil et al, 1991; Anderson et al, 1994; Brettell et al, 1994) will permit the direct testing of such models by the incorporation of the modified genes in wheat and the subsequent commercial availability of wheats with improved dough quality.

REFERENCES

ANDERSON, O.D., BLECHL, A.E., GREENE, F.C., and WEEKS, J.T. 1994. Progress towards genetic engineering of wheat with improved quality. Pages 87-95 in: Improvement of Cereal Quality by Genetic Engineering. R.J. Henry, and J.A. Ronalds, eds. Plenum Press, New York.

ANDREWS, J.L., and SKERRITT, J.H. 1994. Quality related epitopes of wheat high Mr subunits of wheat glutenin. J. Cereal Sci. 19:219-229

BAKER, J.C. and MISE, M.D.1937. Mixing doughs in vacuum and in the presence of various gases. Cereal Chem. 14:721-734.

BELITZ, H.D., KIEFFER, R., KIM, J.J., SEILMEIER, W., and WIESER, H. 1990. Some factors important for gluten functionality. Pages 94-109 in, Interaction of cereal components and their implications for the future. Proc. of ICC-AACC Symposium, Vienna. Eds. R. Lasztity and R.C. Hoseney. ICC. Vienna.

BEKES, F., GRAS,P.W., and GUPTA R.B. 1994a. Mixing properties as a measure of reversible reduction/oxidation of doughs. Cereal Chem 71:44-50.

BEKES, F., ANDERSON, O., GRAS, P.W, GUPTA, R.B., TAM, A., WRIGLEY, C.W., and APPELS, R. 1994b. The contributions to mixing properties of 1D HMW glutenin subunits in a bacterial system. Pages 97-103 in: Improvement of Cereal Quality by Genetic Engineering. Eds. R.J. Henry and J.A. Ronalds. Plenum Publishing Corp., USA.

BEKES, F., GRAS, P.W., GUPTA, R.B., HICKMAN, D.R., and TATHAM, A S. 1994c. Effects of a high Mr glutenin subunit (lBx20) on the dough mixing properties of wheat flour. J. Cereal Sci. 19:3-7.

BEKES, F., and GRAS, P.W. 1992. Demonstration of the 2g Mixograph as a research tool. Cereal Chem 69:229-230.

BELTON, P.S., COLQUHOUN, I.J., FIELD, J.M., GRANT, A., SHEWRY, P.R., and TATHAM, A.S. 1994. ^{1}H and ^{2}H NMR relaxation studies of a high Mr subunit of wheat glutenin and comparison with elastin. J. Cereal Sci. 19:115-121.

BLUMENTHAL, C.S., BARLOW, E.W.R. and WRIGLEY, C.W. 1993. Growth environment and wheat quality: the effect of heat stress on dough properties and gluten proteins. J. Cereal Sci. 18:3-21.

BRETTELL, R.I.S., CHAMBERLAIN, D.A., DREW, A.M., MCELROY, D., WITRZENS, B., and DENNIS, E.S. 1994. Assessment of methods for the genetic transformation of wheat. Pages 3-9 in: Improvement of Cereal Quality by Genetic Engineering. R.J. Henry, and J.A. Ronalds, eds. Plenum Press, New York.

BUSHUK, W., and KAWKA, A. 1990. Chemical reactions and interactions in dough. Pages 8-28 in: Interaction of cereal components and their implications for the future. Proc. of ICC-AACC Symposium, Vienna. Eds. R. Lasztity and R.C. Hoseney. ICC. Vienna.

EWART, J.A.D. 1968. Action of glutaraldehyde, nitrous acid or chlorine on wheat proteins. J. Sci. Food Agric. 19:370-373.
EWART, J.A.D. 1979. Glutenin structure. J. Sci Food Agric. 30:482-492.
EWART, J.A.D. 1988. Studies on disulfide bonds in glutenin. Cereal Chem. 65:95-100.
FENN, D., LUKOW, O.M., BUSHUK, W., and DEPAUW, R.M. 1994. Milling and baking quality of 1BL/1RS translocation wheats. I. Effects of genotype and environment. Cereal Chem.71:189-195.
FIELD, J.M., SHEWRY, P.R., BURGESS, S.R., FORDE, J., PARMAT, S. and MIFLIN, B.J. 1983. The presence of high molecular weight aggregates in the protein bodies of developing endosperms of wheat and other cereals. J. Cereal Sci. 1:33-41.
FIELD, J.M., TATHAM, A.S., and SHEWRY, P.R. 1987. The structure of a high Mr subunit of durum wheat (T. durum) gluten. Biochem. J. 247:215-221.
FINNEY, K.F., and BARMORE, M.A. 1948. Loaf volume and protein content of hard winter and spring wheats. Cereal Chem. 25:291-312
GAO, L., NG, P.K.W., and BUSHUK, W. 1992. Structure of glutenin based on farinograph and electrophoretic results. Cereal Chem. 69:452-455.
GRAS, P.W., BASON, M.L., and TOMLINSON, J.D. 1994. The effect of storage and thermal treatment on the quality of rain-damaged wheat. In: Stored Product Protection. E. Highley, ed. CAB International.Pp.1235-1237.
GRAVELAND, A., BOSVELD, P., LICHTENDONK,W.J., and MOONEN, J.H.E. 1980. Superoxide involvement in the reduction of disulfide bonds of wheat gel proteins. Biochem. Biophys. Res. Commun. 93:1189-1195.
GRAVELAND, A., BOSVELD, P., LICHTENDONK, W.J., MOONEN, J.H.E., and SCHEEPSTRA, A. 1982. Extraction and fractionation of wheat flour proteins. J. Sci. Food Agric. 33:1117-1128.
GRAVELAND, A., BOSVELD, P., LICHTENDONK, W.J., MARSEILLE, J.P., MOONEN, J.H.E., and SCHEEPSTRA, A. 1985. A model for the molecular structure of the glutenins from wheat flour. J. Cereal Sci. 3:1-16.
GUPTA, R.B., and MACRITCHIE, F. 1994. Allelic variation at glutenin subunit and gliadin loci, Glu-1, Glu-3 and Gli-1 of common wheats. II. Biochemical basis of allelic effects on dough properties. J. Cereal Sci. 19:19-29.
GUPTA, R.B., BEKES, F., and WRIGLEY, C.W. 1991. Prediction of physical dough properties from glutenin subunit composition in bread wheats: correlation study. Cereal Chem. 68:328-333.
GUPTA, R.B., KHAN, K, and MACRITCHIE, F. 1993. Biochemical basis of flour properties in bread wheats. 1. Effects of variation in the quantity and size distribution of polymeric protein. J. Cereal Sci. 18:23-41.
GUPTA, R.B. and SHEPHERD, KW. 1990. Two-step one-dimensional SDS-PAGE analysis of LMW subunits of glutenin. I. Variation and genetic control of the subunits in hexaploid wheats. Theor. Appl. Genet. 80:65-74.

HOSENEY, R.C. 1986. Principles of Cereal Science and Technology. Amer. Assoc. Cereal Chemists, St Paul, MN USA.
HOSENEY, R.C and BROWN, R.A. 1983. Mixograph studies V. Effect of pH. Cereal Chem. 60:124-126.
JOLLY, C.J., RAHMAN, S., KORTT, A.A., and HIGGINS, T.J.V. 1993. Characterisation of wheat Mr 15 000 grain softness protein and analysis of the relationship between its accumulation in the whole seed grain softness. Theor. Appl. Genet. 86:589- 597.
JONES, I.K. PHILLIPS, J.W. and HIRD, F.J.R. 1974. The estimation of rheologically important thiol and disulfide groups in dough. J. Sci. Food Agric. 25:1-10.
KASARDA, D.D. 1989. Glutenin structure in relation to wheat quality. Pages 277-302 in: Wheat is Unique. Y. Pomeranz, Ed. Amer. Assoc. Cereal Chem., St Paul, MN.
KASARDA, D.D., KING, G., and KUMOSINSKI, T.F. 1994. Comparison of spiral structures in wheat high-molecular-weight glutenin subunits and elastin by molecular modeling. In: Molecular modeling. T.F. Kumosinski, and M. Liebman, eds. ACS Monograph, Amer. Chem. Soc. Pp. 209-220.
KASARDA, D.D., NIMMO, C.C., and KOHLER, G.O. 1978. Proteins and the amino acid composition of wheat fractions. Adv. Cereal Sci. Technol. 1 :227-299.
KAWAMURA, Y., MATSUMURA, Y., MATOBA, T., YONEZAWA, D, and KITO, M. 1985. Selective reduction of interpolypeptide and intrapolypeptide disulfide bonds of wheat glutenin from defatted flour. Cereal Chem. 62:279-283.
KECK, B., KOHLER, P., and WIESER, H. 1995. Disulfide bonds in wheat gluten.Peptic and Thermolytic cystine peptides from gluten proteins. Z. Lebensm. Unters. Forsch. 200: 432-439
KHAN K., and BUSHUK, W. 1979. Studies of glutenin. XII. Comparison by sodium dodecyl sulfate-polyacrylamide gel electrophoresis of unreduced and reduced glutenin from various isolation and purification procedures. Cereal Chem. 56:63-68.
KHAN, K, and HUCKLE, L 1992. Use of multistacking gels in sodium dodecyl sulfate- polyacrylamide gel electrophoresis to reveal polydispersity, aggregation, and disaggregation of the glutenin protein fraction. Cereal Chem. 69:686-8.
KOHLER, P., BELITZ, H.D., and WIESER, H. 1991. Disulphide bonds in wheat gluten: isolation of a cystine peptide from glutenin. Z. Lebensm. Unters. Forsch. 192:234-239.
KOHLER, P., BELITZ, H.D., and WIESER, H. 1993. Disulphide bonds in wheat gluten: further cysteine peptides from high molecular weight (HMW) and low molecular weight (LMW) subunits of glutenin and from gamma-gliadins. Z. Lebensm. Unters. Forsch. 196:239-247.

KRETOVICH, V.L., and VAKAR, A.B. 1964. Effect of D_20 on the physical properties of wheat gluten. Proc. Acad. Sci. USSR, Sect. Biochem. 155:71.
LAWRENCE, G.J., and PAYNE, P.I. 1983. Detection by gel electrophoresis of oligomers formed by the association of high molecular weight glutenin subunits of wheat (Triticum aestivum) endosperm. J. Exp. Bot. 34:254-267.
LEW, E.J.L, KUZMICKY, D.D., and KASARDA, D.D. 1992. Characterisation of low molecular weight glutenin subunits by reversed-phase high-performance liquid chromatography, sodium dodecyl sulfate-polyacrylamide gel electrophoresis, and N- terminal amino acid sequencing. Cereal Chem. 69:508-515.
MACRITCHIE, F 1992. Physicochemical properties of wheat proteins in relation to functionality. Adv. Food Nutr. Research 36:1- 87.
MACRITCHIE, F, DUCROS, D.L, and WRIGLEY, C.W. 1990. Flour polypeptides related to wheat quality. Adv. Cereal Sci. Technol. 10:79-145.
MACRITCHIE, F., and GUPTA, R.B. 1993. Functionality composition relationships of wheat flour as a result of variation in sulfur availability. Aust. J. Agric. Res. 44:1767-1774.
METAKOVSKY, E.V., NG,.P.K.W., CHERNAKOV, V.M., POGNA, N.E., and BUSHUK,W. 1993. Gliadin alleles in Canadian western red spring wheat cultivars: use of two different procedures of acid polyacrylamide gel electrophoresis for gliadin separation. Genome 36:743-749.
MILES, M.J., CARR, H.J., McMASTER, T.C., I'ANSON, K.J., BELTON, P.S., MORRIS, V.J., FIELD, J.M., SHEWRY, P.R. and TATHAM, A.S. 1991. Scanning tunneling microscopy of a wheat seed storage protein reveals details of an unusual supersecondary structure. Proc. Natl. Acad. Sci. 88: 68-71.
MOONEN, J.H.E., SCHEEPSTRA, A. and GRAVELAND, A. 1985. Biochemical properties of some high-molecular weight glutenin subunits of wheat glutenin. J. Cereal Sci. 3:17-27.
MORRISON, W.R., GREENWELL, P., LAW, C.N., and SULAIMAN, B.D. 1992. Occurrence of friabilin, a low molecular weight protein associated with grain softness, on starch granules isolated from some wheats and related species. J. Cereal Sci. 15:143- 149.
MULLER, S., and WIESER, H. 1995. The location of disulfide bonds in alpha-type gliadins. J. Cereal Sci. 22:21-28.
NG, P.KW., XU, C. and BUSHUK, W. 1991. Model of glutenin structure based on Farinograph and electrophoretic results. Cereal Chem. 68:321-322.
O'BRIEN, L., and ORTH, R.A. 1974. Effect of geographic location of growth on wheat milling yield, Farinograph properties, flour protein and residue protein. Aust. J. Agric. Res. 28:5-9.
PANOZZO, J.F., EAGLES, H.A., BEKES, F., and WOOTTON, M. 1994a. Genotype and environment interactions on wheat quality in Australia. In: Proc 44th Australian Cereal Chemistry Conference. J.F. Panozzo, ed. Royal Aust. Chem. Inst., Melbourne (in press).

PANOZZO, J.F., BEKES, F., WRIGLEY, C.W. and GUPTA, R.B. 1994b. The effects of bromate (0-30ppm) on the proteins and lipids of dough. Cereal Chem. 71:195-199.
PAYNE, P.I. 1987. Genetics of wheat storage proteins and the effect of allelic variation on bread- making quality. Ann. Rev. Plant Physiol. 38:141-153.
PRITCHARD, P.E., and BROCK, C.J. 1994. The glutenin fraction of wheat protein: the importance of genetic background on its quantity and quality. J. Sci. Food Agric. 65:401-406.
POGNA, N., LAFIANDRA, D., FEILLET, P., and AUTRAN, J.C. 1988. Evidence for a causal effect of low molecular weight subunits on gluten viscoelasticity in durum wheats. J. Cereal Sci. 7:211-214.
RATH C.R., GRAS, P.W., WRIGLEY C.W., and WALKER, C.E. 1990. Evaluation of dough properties from two grams of flour using the Mixograph principle. Cereal Foods World 35:572- 574.
RANDALL, P.J., and WRIGLEY, C.W. 1986. Effects of sulfur supply on the yield, composition and quality of grain from cereals, oilseeds and legumes. Adv. Cereal Sci. Technol. 8:171-206.
SCHNUG, E., HANEKLAUS, S., and MURPHY, D. 1993. Impact of sulphur supply on the baking quality of wheat. In: Aspects of Applied Biology 36, Cereal Quality III. Eds. Kettlewell, P.S., Garstang, J.R., Duffus, C.M., Magan, N., Thomas, W.T.B., and Paveley, N.D. Assoc. Appl. Biologists, Warwick, U.K.
SHANI, N., ROSENBERG, N., KASARDA, D.D., and GALILI. 1994. Mechanisms of assembly of wheat high molecular weight glutenins inferred from expression of wild-type and mutant subunits in transgenic tobacco. J. Biol. Chem. 269:8924- 8930.
SHANI, N., STEFFEN-CAMPBELL, J.D., ANDERSON, O.D., GREENE, F.C. and GALILI, G. 1992. Role of the amino- and carboxy-terminal regions in the folding and oligomerization of wheat high molecular weight glutenin subunits. Plant Physiol. 98:433- 441.
SHEPHERD, K.W., SINGH, N.K., GUPTA, R.B., and KOEBNER, R.M.D. 1991. Quality characteristics of flour from wheat-rye translocation and recombinant lines Pages 715-723 in, Gluten proteins 1990 Eds. W. Bushuk and R. Tkachuk, Am. Assoc. Cereal Chem., St Paul, Minn., USA.
SIMMONDS, D.H., WRIGLEY, C.W., and GRAS, P.W. 1972. Treatment of flour to control physical properties of doughs made therefrom. Aust. Patent Appl. PA7902.
SWALLOW, W.H., and EVERY, D. 1991. Insect enzyme damage to wheat. Cereal Foods World 36:505-508.
TANAKA, K, and BUSHUK, W. 1973. Changes in flour proteins during dough-mixing. II. Get filtration and electrophoresis results. Cereal Chem. 50:597-605.

TAO, H.P., ADALSTEINS, A.E., and KASARDA, D.D. 1992. Intermolecular disulElde bonds link specific high-molecular-weight glutenin subunits in wheat endosperm. Biochim. Biophys. Acta. 1156:13-21.
VASIL, V., BROWN, S.M., RE, D., FROMM, M.E., and VASIL, I.K 1991. Stably transformed callus lines from micro-projectile bombardment of cell suspension cultures of wheat. Bio/Technology 9:743-747.
WEEGELS, P.L., DE GROOT, A.M.G., VERHOEK, J.A., and HAMER, R.J. 1994. Effects on gluten of heating at different moisture contents. II Changes in physico-chemical properties and secondary structure. J. Cereal Sci. 19:39-47.
WERNER, W.E. and KASARDA, D.D. 1991. Composition of HMW-glutenin subunit dimers formed by partial reduction of residue glutenin. Cereal Foods World 36:722 (Abstract).
WIESER, H., SEILMEIER, W., and BELITZ, H.D. 1994. Quantitative determination of gliadin subgroups from different wheat cultivars. J. Cereal Sci. 19:149-155.
WRIGLEY, C.W. 1994 Developing better strategies to improve grain quality for wheat and barley. Aust. J. Agric. Research 45:1- 17.
WRIGLEY, C.W., LEE, J.W., GRAS, P.W., and MACRITCHIE, F. 1972. The mechanical properties of dough in relation to the molecular weight distribution of gluten proteins. Proc. 22nd Aust. Cereal Chem. Conf. Pp. 25-32. Royal Aust. Chem. Inst., Melbourne.
WRIGLEY, C.W., GUPTA, R.B. and BEKES, F. 1993. Our obsession with high resolution in gel electrophoresis; does it necessarily give the right answer? Electrophoresis 14:1257-1258.

CHAPTER 3
LIPID-CARBOHYDRATE INTERACTIONS

Ann-Charlotte Eliasson,
Dept of Food Technology, University of Lund,
PO Box 124, S-221 00 LUND, Sweden

INTRODUCTION

Lipids as well as carbohydrates are present in all cereals, and it is thus to be expected that their interactions will be of importance for product quality. This chapter explores what kind of interactions are possible, and how these interactions exhibit themselves during processing and in products.

Lipids and carbohydrates are not in direct contact in the cereal grain, because of the compartmentalization of the kernel. One exception is the amylose-lipid complex, that recently has been shown to exist in the native starch granule (Morrison et al, 1993). The compartmentalization will break up during milling, but there will still not be any interactions between carbohydrates and lipids in the flour. Interactions would not be expected to take place until water is added, and the flour is processed to a food product. For lipid and starch to interact to a measurable degree also an increase in temperature is usually necessary. In many products, lipids (emulsifiers or shortenings) and/or carbohydrates (sucrose or polysaccharides) might be added, thus giving new opportunities for interactions. The description in the present chapter starts with general knowledge about carbohydrate-lipid interactions and the knowledge about what carbohydrates and lipids that are present in cereals and cereal products. The main focus will then be on starch-lipid interactions, and their influence on product quality will be described, as well as possible mechanisms for the interaction. The formation of the amylose-lipid complex is probably the most important mechanism, and this complex will be described in some detail.

SOME GENERAL ASPECTS ON LIPID-CARBOHYDRATE INTERACTIONS

To discuss the interactions between lipids and carbohydrates in cereal products a good starting point could be to briefly outline what kind of interactions that are known to take place between lipids and carbohydrates.
In this chapter "interactions" are defined as non-covalent interactions, such as ionic, hydrophobic, or hydrogen bonding. For the discussion of possible interactions it is convenient to divide the carbohydrates in two groups according to their molecular weight. The low-molecular weight carbohydrates (mono-, di- and oligosaccharides) exert their influence mainly through their

solubility. The polysaccharides, being polymers, are of much greater interest in relation to interactions with lipids. Parallels can be made with non-food systems containing polymers and surfactants (Goddard, 1993a).

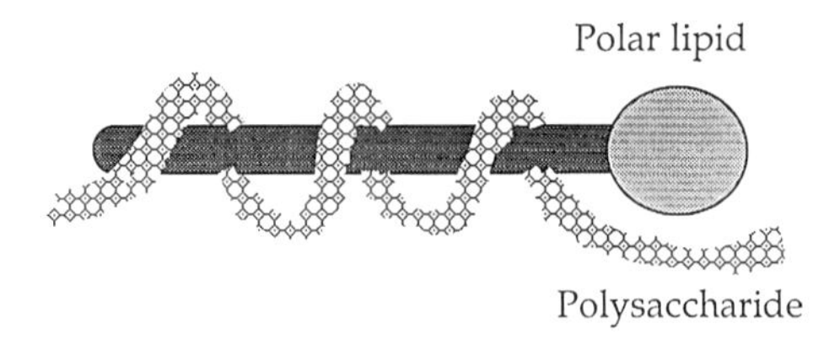

Figure 1. An example of binding of a lipid molecule (e.g. monoglyceride) to a polysaccharide (e.g. amylose).

In general, when lipids and polysaccharides are mixed, not very much happens, except in some special cases when there is an interaction. It is possible to identify two types of interactions: the molecular "binding" of a lipid molecule to the polysaccharide (Fig. 1), and the interaction between a lipid phase (micelles) and the polysaccharide (Fig. 2).

Figure 2. An illustration of the interaction between a lipid phase (e.g. SDS-micelles) and a polysaccharides (e.g. a cellulose derivative).

The polysaccharide-lipid interaction illustrated in Fig. 1 is perhaps best exemplified with the well-known amylose-lipid complex. In this case individual lipid molecules interact with the amylose molecule. The main driving force for complex-formation is the hydrophobic effect, i.e. the amylose helix offers a better environment for the hydrocarbon chain than the aqueous surrounding.

The amylose-lipid complex is the lipid-carbohydrate interaction that has been most studied for food systems, and this molecular interaction is usually used to explain different phenomena, involving starch and lipids, in food processes or food products. Examples are the addition of lipids to decrease stickiness of pasta, mashed potatoes and candies, and to improve shelflife of bread. As will be evident in this chapter not all these effects can be explained with the commonly accepted amylose-lipid complex, and therefore other types of interactions have to be considered as well.

Fig. 2 illustrates another type of interaction, where surfactant micelles interact with the polymer. For any surfactant there is a critical concentration, the critical micelle concentration (c.m.c.), when micelles start to form. In the presence of a polymer the micelles form at a lower concentration, the critical aggregate concentration (c.a.c.) (Goddard, 1993a). The main driving force in this interaction is the hydrophobic interaction among the surfactant chains (Lindman and Thalberg, 1993).

This kind of interaction has been investigated for a range of systems, although so far not in food systems. One of the most well characterised systems is poly(ethylene oxide) and sodium dodecylsulphate, SDS (Lindman and Thalberg, 1993). Another example is sodium hyaluronate and tetradecyltrimethyl-ammoniumbromide (Thalberg et al., 1990). An example with direct relevance for cereal products is the interaction between cellulose derivatives and surfactants (Carlsson et al., 1988). One interesting feature of these systems is their gelling ability, something that has attracted interest for technical applications in, for example, the pharmaceutical industry (Lindman and Thalberg, 1993). The polymer-surfactant interactions can be used for rheology control, viscosity enhancement, gel formation, and solubilization, but they also have implications for surface activity and adsorption (Goddard, 1993b).

IMPORTANCE OF LIPID PROPERTIES FOR THE INTERACTION

In this section some physical properties of the lipid that are of importance for interactions with carbohydrates will be described. The word "lipid" is used to denote native lipids as well as emulsifiers or surfactants that might be added. In case of the latter group, no distinction is made between emulsifiers or surfactants extracted from biological materials or chemically produced. For the discussion of lipid-carbohydrate interactions it is the functional properties of the compounds that are most relevant, not their origin. It is therefore useful to divided the lipids into two groups, based on their interaction with water: polar lipids and non-polar lipids.

The non-polar lipids do not interact with water, and a familiar representative for this group is the triglyceride, i.e. fats, oils, and shortenings. The triglycerides have the ability to crystallise, and for most fats three polymorphic forms are discernible: α, β' and β (Larsson, 1986). Baking results depends on the polymorphic form of the added fat. For cakes β' seems to be the most effective (Krog et al., 1985), whereas in bread making the β'-form has the largest influence on loaf volume (Leissner, 1986). Another important property of triglycerides that may influence interactions with carbohydrates is the melting temperature. Fats, shortenings, and margarines used in baking are composed of a mixture of triglycerides with different melting temperatures. The melting temperature of the fat will thus not be a single well-defined temperature, but rather a temperature range. At a temperature melted liquid fat, and a proportion of crystalline, solid fat. The polar lipids, including emulsifiers, interact with water to form liquid-crystalline phases (Larsson, 1986). Phase diagrams, showing the relation between concentration, temperature, and the phase behaviour, have been published for several commercial emulsifiers (cf. Krog et al (1985)).

Phase diagrams have also been published for wheat and other cereal lipids (Carlson et al, 1978; Carlson et al, 1979b; Carlson et al, 1980). For the interaction with carbohydrates the phase behaviour of the lipid is of importance, and it might influence the interaction in several ways. For a molecular interaction, the monomer concentration is of course important. For a monoglyceride such as mono-olein association to liquid-crystalline phases starts at concentrations above about 1 x $10^{-6\,m}$ (Krog et al, 1985). This could be compared with the c.m.c., which is around 1 mM for many surfactants (Mukerjee and Mysels, 1971). The lipid phase present influences the distribution of the lipid in the aqueous food system. It is, for example, much easier to disperse the lamellar liquid-crystalline phase than the cubic phase, because in the lamellar phase water is the continuous phase. The different liquid-crystalline phases have different rheological properties, and the cubic phase is highly viscous (Krog et al, 1985). This further reduces the possibility to distribute the lipids when present in the cubic phase. Thus, if a polar lipid is added to reduce stickiness by complex-formation with amylose the effect may completely come off if the lipid is added at conditions resulting in the formation of the cubic liquid-crystalline phase.

The phase behaviour will determine how effective an added emulsifier will be. This has been convincingly illustrated in baking experiments, where lipids added in the lamellar liquid-crystalline phase give the best loaf volume (Krog and Nybo-Jensen, 1970; Carlson, 1981; Rajapaksa et al, 1983). The ability of an emulsifier in delaying the staling of bread is also related to the phase behaviour of the additive.

Another important difference between fats and emulsifiers is the amounts that are used in cereal products. Fats are usually added at much higher levels than emulsifiers, and the proportion of fat in cakes can be perhaps 50% (based on the flour weight), whereas in bread maybe 0.5% of an emulsifier is added.

Polar lipids differ in their ability to participate in the molecular interaction illustrated in Fig. 1. This has been studied in great detail for the amylose-lipid complex (Krog, 1971). Complex formation is found to be influenced by the polar head as well as by the hydrocarbon chain length (Kowblansky, 1985; Eliasson, 1986a). The best complex forming ability is found for chain lengths around C14-16, with non-ionic heads (Krog, 1971). The complex forming ability is strongly related to the phase behaviour. When unsaturated monoglycerides were added to an amylose solution at room temperature a very small amount of complex was formed (Riisom et al, 1984). If, however, the cubic phase formed during these conditions was transformed into liposomes, the complex formation increased.

The type of lipid present in an amylose-lipid complex (Fig.1) will also affect what properties those complexes have, e.g. thermal stability, stability against enzymatic breakdown, gel properties, and others.

LIPID-CARBOHYDRATE INTERACTIONS IN CEREAL PRODUCTS

Lipid-carbohydrate interactions are of importance for the quality of all cereal products, simply because both lipids and carbohydrates are present in all flours. Thus, these interactions constitute an inevitable part of product quality. However, it is impossible to describe what contribution the interaction between the native lipids and carbohydrates make to the final product quality, because it is more or less impossible to prepare a product without these interactions taking place. Defatting of starch, for example, invariably causes such changes in the starch granule that it can no longer be regarded as "native" (Morrison and Coventry, 1985). The influence of lipid-carbohydrate interactions on product quality is far more easily identified when lipids and/or carbohydrates are added to a batter or a dough. Some specific examples of the contributions of lipid-carbohydrate interactions are the incorporation of air in batters because of the mixing of sucrose crystals and solid fat, the addition of emulsifiers to pasta to decrease stickiness, and the addition of emulsifiers to bread to decrease the staling rate. That interactions between components are of importance for product quality becomes evident when one component is replaced, as for example in the making of low fat bakery products (Silverio and Eliasson, 1994). That such interactions are crucial for product quality means that a fat replacer must have the same effect, either by interacting itself, or by achieving the same effect in another way. A different example is the production of gluten-free products for those suffering from the celiac disease. In such products hydrocolloids in combination with emulsifiers might be used as a gluten replacer. To use the polysaccha-ride-surfactant interaction, described in Fig. 2, is perhaps a new direction for developing such products.

To be able to describe the importance of lipid-carbohydrate interactions for product quality it is necessary to know something about what lipids and what carbohydrates are present. The ingredients (including additives), besides the flour, used in bakery products might be of lipid or carbohydrate nature, and thus give new possibilities for lipid-carbohydrate interactions. The amounts and composition of the carbohydrates in some cereals are given in Table I.

Table I shows that polysaccharides are the most abundant of the carbohydrates, and of these, starch is the far most common. Starch is also the carbohydrate of most importance for lipid-carbohydrate interactions. Polysaccharides are used as stabilisers in foods, but this is not so common in cereal products, except, perhaps, for low-fat bakery products. Therefore the only added carbohydrate that will be discussed is sugar (i.e. sucrose)

TABLE I
Composition of carbohydrates in some cereal grains

Component	Wheat[a] (inner endosperm)	Rye[b]	Maize[c]
Sugars/reducing sugars (g/100 g)	1.6	---[d]	2.6
Starch (g/100 g)	72.6	56	72
Pentosans (g/100 g)	1.4	10	---
Cellulose (g/100 g)	0.3	---	---
Fiber (g/100 g)	---	---	9.5

Data from [a]Johnson (1991), [b]Lorenz (1991), [c]Mattern (1991). [d]Dotted lines indicate that data were not given in the references.

Detailed information is available on the lipid composition of cereals (Morrison, 1978). However, for the study of lipid-carbohydrate interactions, such detailed knowledge may not be necessary. As discussed above it is usually enough to divide the lipids in polar lipids and non-polar lipids, because from the functional point of view (for example the phase behaviour) a mixture of polar lipids behaves as one component (Larsson, 1986). Some of the most common lipids in a few cereals are listed in Table II.

The emulsifiers used in cereal products include monoglycerides (MG), diacetyl tartaric acid esters of monoglycerides (DATEM), and sodium stearate lactylate (SSL). Fats, oils, shortenings, and margarines are complicated mixtures of triglycerides, and their composition will not at all be addressed in the present chapter.

TABLE II
The composition of cereal lipids in flour (wheat and rye) and in kernel (maize)[a]

Lipid class	Wheat (mg/100 g d.b.)	Rye (mg/100 g d.b.)	Maize (mg/100 g d.b.)
Steryl ester	72	226	138
Triglyceride	674	422	12,751
Diglyceride	86	163	152[b]
Monoglyceride	66	61	---
Free fatty acid	110	130	41
Galactosyl compounds	382	123	---
Ethanolamine compounds	106	---	---
Phosphatidyl compounds	90	144	---
Lysophosphatidyl compounds	46	44	---
Total non-polar	1079	1007	---
Total glycolipids	382	140	---
Total phospholipid	242	159	---
Total polar	---	---	538

[a]Data calculated from Morrison (1978). [b]Including monoglycerides

SUGAR-LIPID INTERACTIONS

Interactions between low-molecular weight carbohydrates, sugars, and lipids are known to take place in at least one system - cell membranes during drying. Organisms that can survive drying, and regain activity when rehydrated, contain large amounts of trehalose. Sucrose may in some organisms work in the same way as trehalose. To explain this phenomenon a "water replacement" hypothesis has been suggested (Crowe et al, 1988). The structure of proteins and the assembly of phospholipids in cell membranes depend on the presence of non-polar regions of the molecules that try to avoid water contact. When water is removed the need to avoid contact with water is also removed and structures collapse. If trehalose is present it can replace water and interact with, for example, the polar head of a phospholipid, and the structure remains.

Interactions between sugars and lipids have been observed in several model systems that may be relevant for cereal products. When vesicles and liposomes of phospholipids are freeze-dried in the presence of trehalose, and then rehydrated, no change in properties is observed (Crowe et al, 1988). Freeze-drying in the absence of sugars causes the liposomes or vesicles to collapse. Thus, in the presence of trehalose the lamellar liquid-crystalline phase is preserved in the dry state, where otherwise (no sugar) the gel state would have been obtained.

Moreover, drying is not mandatory for sugar-lipid interactions to take place. When water in an aqueous monoglyceride system is replaced with a sucrose solution the phase behaviour of the lipid changes: there will be a transition from the cubic phase to a hexagonal phase (Soderberg and Ljusberg-Wahren, 1990). Sucrose is more effective than fructose in promoting this transition. This phenomena certainly has applications for cakes and other products with a high sucrose content and where added emulsifiers are used for stabilising a whipped foam. Thus, the phase-behaviour of an added emulsifier might differ from what could be expected from a phase-diagram for the actual lipid-water system, and the selected emulsifier may not work as an emulsifier in the actual recipe.

In a whipped batter the emulsifier is added to promote the incorporation of air into the aqueous batter, whereas during mixing of a solid fat and sugar air is incorporated into the fat phase. In the latter case the sugar crystals are important, because their irregular shape will pull air with them into the fat. During mixing fat surrounds the air cells, and thus incorporate them into the batter. This is an example of an interaction on the colloidal, or even microscopic, level rather than on the molecular level.

PENTOSAN-LIPID INTERACTIONS

Pentosans, cell-wall components in cereals, are present in cereal flours in different amounts depending on the cereal (Table I). Another cell-wall component is the β-glucan, typical of barley and oats. The pentosans are often divided into water-soluble and water-insoluble fractions. The main components in the pentosan group are the arabinogalactans and the arabinoxylans. The protein content in different pentosan fractions is much discussed in the literature, and it is speculated that some fractions could be glycoproteins (cf. Eliasson and Larsson (1993).

Pentosan-lipid interactions, or interactions between lipids and other cell wall components, have so far been neglected. However, there is at least one field in which this type of interaction might be of significance. It has recently been shown that arabinoxylans and arabinogalactans from wheat are surface active (Izydorczyk et al, 1991). Measurements in our laboratory have confirmed that arabinoxylans from rye are surface active (Wannerberger, L., unpublished results). Arabinoxylans from wheat has the ability to stabilise protein foams after heating, a property that the arabinogalactans lack

(Izydorczyk et al, 1991; Izydorczyk and Biliaderis, 1992). There are thus three different groups of compounds that might compete for the air/water interface in doughs: proteins, polar lipids and arabinoxylans. It has been proposed that the proteins are of importance for stabilising the air/water interface during mixing and fermentation, whereas the lipids stabilise the interface during the early stages of oven spring (Eliasson and Larsson, 1993). Mixed films of protein and lipids show poor rheological properties (Paternotte et al, 1994), and could explain the poor baking result obtained when low levels of lipids are added to a defatted flour (MacRitchie and Gras, 1973). The composition of the air/water interface in dough as well as during the early stages of baking might be crucial for the baking performance, and pentosan-lipid interactions should be studied in this respect.

CELLULOSE-LIPID INTERACTIONS

Cellulose is present at rather low levels in cereal flours (see Table I), but it might also be added to increase the dietary fibre content of, e.g., wheat flours (Pomeranz et al, 1977). Addition of fibers (or non-wheat flours) has a deteriorating effect on baking performance and bread quality. This might be overcome with the addition of polar lipids in the lamellar liquid-crystalline phase (Rajapaksa et al, 1983). However, the improving effect is because of stabilising the air/water interface rather than a cellulose-lipid interaction.

To find information about cellulose-lipid interactions it is necessary to look outside the cereal field. As already noted, interactions between cellulose derivatives and surfactants have achieved considerable attention because of the gelling ability. When a solution of ethyl(hydoxyethyl)cellulose (EHEC) was heated the viscosity, as expected, decreased (Carlsson et al, 1990). However, when a surfactant (CTAB; cetyltrimethylammonium bromide) was present the viscosity increased with temperature, and a gel formed. With further heating the viscosity decreased. Also for cationic cellulose and SDS, gels are formed (Goddard, 1993a). It is generally found that gel formation occurs under particular conditions of temperature and certain concentration of surfactants. The gel forming mechanism could be a combination of increased entanglements, and the association of a surfactant aggregate to different polymer chains. Each micelle, illustrated in Fig. 2, could be depicted to "bind" to two different polymer chains.

STARCH-LIPID INTERACTIONS

Lipids and starch are both present in any cereal flour, and their interaction will constitute a necessary part of the process converting raw material to the end product. As already discussed, the lipids are impossible to extract, without destroying the starch granule. It is thus impossible to obtain a product prepared without starch-lipid interactions. To elucidate the influence of starch-lipid interactions on product quality more lipids can be added, and the effects studied. This can then be extrapolated back to conditions without lipids. For some starches, e.g. oats (Gudmundsson and Eliasson, 1989), it is possible to compare the behaviour of a range of starches differing in lipid content. Comparisons between cereal starches and potato starch, a starch essentially free of lipids, can also give information about the significance of starch-lipid interactions. The influence of the interactions on starch behaviour as well as on lipid behaviour will be described.

Starch-lipid interactions take place in at least three different ways. Probably the most common and well-known is the formation of the amylose-lipid complex (French and Murphy, 1977). Recently, a complex formation between polar lipids and the outer branches of amylopectin has been suggested (Evans, 1986; Slade and Levine, 1987; Eliasson and Ljunger, 1988; Gudmundsson and Eliasson, 1992). A third possibility could be the adsorption of micelles or lipid phases (Larsson, 1983) onto the starch granules or starch polymers. These and other starch-lipid interactions will be discussed in the final part of this section.

Gelatinization

When starch is heated in water gelatinization occurs, and gelatinization, at least to some extent, is essential in most products. Examples are the setting of the crumb in bread and cakes, and thickening of the stuffing in other products. Thus it can be assumed that an interference with the gelatinization process will greatly affect the product quality. Gelatinization involves several processes, including loss of birefringence and crystallinity, absorption of water, swelling, solubilization of amylose/amylopectin, and formation of a paste or a gel (or at least an increase in viscosity). The course of events can be followed using several methods including rheological measurements, X-ray diffraction techniques, microscopy, and differential scanning calorimetry (DSC). The same methods have been used to study the influence of lipids on the behaviour of starch.

The gelatinization of wheat starch has been studied by DSC for extracted starch and for starch in wheat flours (Eliasson, 1989). Even in case of wheat flour it is the starch component that is measured, because the denaturation of wheat proteins causes insignificant endotherms in the DSC- thermogram (Eliasson and Hegg, 1981). An obvious result is that the enthalpy due to the transition of the amylose-lipid complex is higher in flour than in starch (1.8 J/g dry matter compared with 1.5 J/g dry matter). Although amylose-lipid complexes are known to exist in the native starch granule (Morrison et al., 1993), the presence of flour lipids leads to the formation of new complexes.

The gelatinization onset temperature (T_0) and the temperature at peak maximum (T_m) in the DSC-thermogram was found to be higher in flour than in starch. This is consistent with the findings when polar lipids have been added to starch (see below). When a range of different wheat flours were compared (including spring, winter, feed, biscuit, and durum wheats) no significant differences in the transition endotherm of the amylose-lipid complex were found (Eliasson et al., 1995)

When certain emulsifiers are added a slight increase in T_o has been observed, as illustrated in Table III. The effect is, however, rather small, only 1- 2°C. For the gelatinization enthalpy (ΔH) a decrease is usually observed, probably related to exothermic complex formation occurring at the same temperature range as the gelatinization (Evans, 1986). A slight increase in the temperature for the typical morphological changes of wheat starch during heating has been observed in the presence of emulsifiers (Eliasson, 1985a). It thus seems that the main effect of polar lipids on starch gelatinization is to cause a slight increase in temperature. However, it should be mentioned that there is at least one exception from this, namely SDS. This surfactant decreases the gelatinization temperature (Eliasson, 1985a; Gough et al, 1985). The same effect is observed for lysolecithin, the native lipid present in wheat starch, but not for cationic surfactants (Table III).

TABLE III
Change in gelatinization onset temperature (T_o), maximum temperature (T_m) and enthalpy (ΔH) of wheat starch when emulsifiers and triglycerides are added[a].

Additive[b]	T_o (°C)	T_m (°C)	ΔH (J/g d.b.)
Lysolecithin[c]	-1.3	-0.7	-5.2
SSL[c]	1.4	1.0	-2.4
CTAB[c]	0.6	0.4	-4.2
SDS[c]	-2.3	-1.2	-3.2
Lecithin[c]	-0.3	-0.5	-0.5
Triglycerides[d]	-0.4	-0.3	0.1

[a]Starch : water = 1:3 + 5 % emulsifier, starch : water = 1 : 1 + 50 % triglycerides.
[b]SSL = sodium stearate lactylate, CTAB = cetyltrimethylammoniumbromide, SDS = sodiumdodecylsulphate.
[c]Data from Eliasson (1986a).
[d]Data from Silverio and Eliasson (1994).

During the oven step in the breadmaking process the crumb structure is set as a result of starch gelatinization. This process puts an end to the volume expansion of the dough. A higher temperature for the onset of gelatinization thus gives a somewhat longer period for volume increase, and a relation between loaf volume and starch gelatinization temperature has, in fact, been found (Soulaka and Morrison, 1985). It could be assumed that one of the functions of emulsifiers is to delay the setting of the crumb, and thus allow a prolonged time for volume expansion (cf. Junge et al (1981)).

When triglycerides are added to starch no, or very small, effects on the gelatinization process have been observed (Table III). It should be noted when the results in Table III are compared, that much higher levels of triglycerides are added than of the emulsifiers. One very important conclusion from this result is that fats do not interfere with the gelatinization of starch, and, as should be expected, fats do not influence the water distribution in the system. This behaviour has implications for the use of fat replacers in bakery products with a reduced fat content. For a fat replacer to fulfil the role of fat it is essential that no interaction with the water takes place (van Gijssel and Eliasson, 1993; Silverio and Eliasson, 1994). During gelatinization, amylose leaches out of starch granules, and depending on the conditions amylopectin might also be leached. When lipids are present, e.g. an added emulsifier, the leaching of amylose is decreased (Fig. 3). Stickiness, a result of the leached amylose, decreases when emulsifiers are added. When several cereal starches were compared it was found that the lowest amylose leaching was obtained for oat starches (Gudmundsson, 1992a). For this species the amylose leaching decreased with increasing lipid content of the starch.

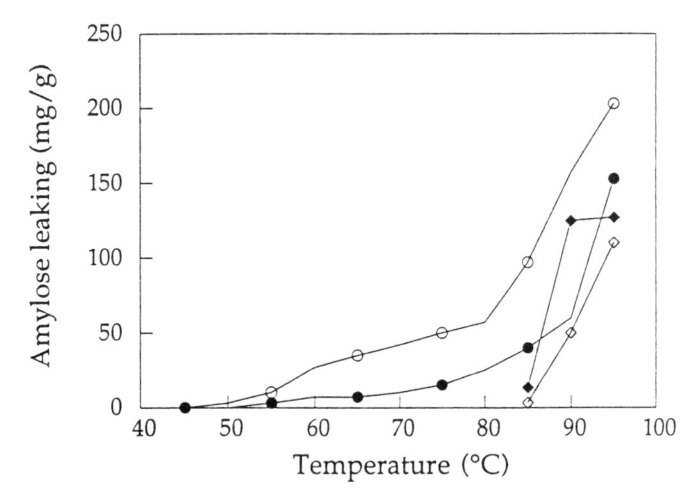

Figure 3. The influence of added lipids on the leaching of amylose during gelatinization, O = wheat starch without additive, ● = wheat starch with monoglycerides, ◇ = oat starch with low lipid content, ♦ = oat starch with high lipid content. Data taken from Eliasson (1985a) and Gudmundsson and Eliasson (1989).

The presence of lipids, resulting in a decreased amylose leaching, has implications for the formation of an amylose gel. Less amylose will be available for the formation of a network, and a softer product could be expected when lipids are added. Amylose is suggested to play an important role for the crumb structure of the newly baked bread (Ghiasi et al, 1984), and the addition of emulsifiers could be expected to change the crumb structure. Initially softer breads are obtained with some added emulsifiers, (Krog et al, 1989), but as will be discussed later, lipids can influence the rheological properties of starch in many different ways.

At the same time as the leaking of amylose occurs, the starch granules swell. The swelling is affected by lipids present, and the native lipids decrease or delay the swelling as illustrated for oats in Fig. 4. Most added emulsifiers decrease the swelling, as illustrated for the monoglyceride in Fig. 4. Again, SDS is an exception, and swelling occurs at lower temperatures when this surfactant is added (Eliasson, 1985a).

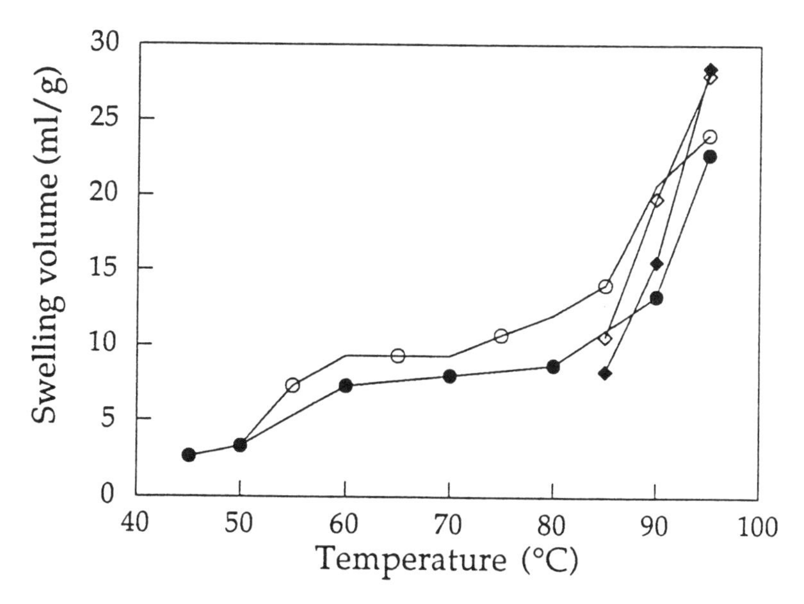

Figure 4. The influence of added lipids on the swelling of starch during gelatinization, O = wheat starch without additive, ● = wheat starch with monoglycerides, ◇ = oat starch with low lipid content, ♦ = oat starch with high lipid content. Data taken from Eliasson (1985a) and Gudmundsson and Eliasson (1989).

When bread is stored certain detrimental changes occur that are referred to as staling (Kulp and Ponte, 1981). These changes include an increase in firmness, something that the consumers are well aware of. If emulsifiers are added (e.g. MG or DATEM) the bread will keep soft for a longer period of time (Russell, 1983; Krog et al, 1989). This is because of both an initial greater softness (at least for some emulsifiers), and to a decreased firming rate.

During the same time period as the bread becomes harder an increase in starch crystallisation ("retrogradation") can be measured using X-ray diffraction (Zobel, 1973), or DSC (Russell, 1983; Eliasson, 1985b). Interestingly enough the B-pattern is formed during storage of bread, contrary to the native wheat starch in the dough, which exhibits the A-pattern. If emulsifiers are present the V-pattern will emerge (Zobel, 1973), i.e. the pattern characteristic of the amylose-lipid complex.

When the retrogradation of starch is followed using the DSC it is the crystallisation of amylopectin that is monitored (Eliasson, 1985b). The enthalpy of the melting of crystallised amylopectin (ΔH_c) has been used as a measure of staling (Russell, 1983). An increase in the transition due to the amylose-lipid complex is observed when emulsifiers are present (Russell, 1983; Krog et al, 1989).

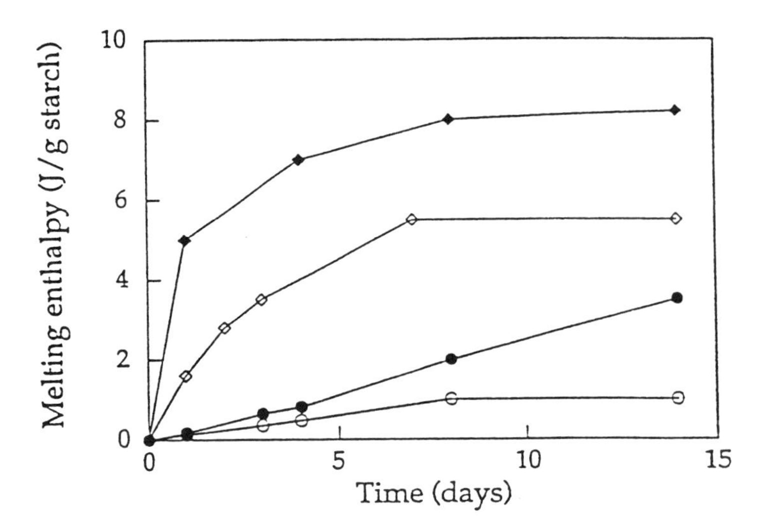

Figure 5. Retrogradation of cereal starches as measured by DSC after storage at room temperature, O = oat starch with high lipid content, ● = oat starch with low lipid content, ◊ = rye starch, ♦ = maize starch. Data taken from Gudmundsson and Eliasson (1989) and Gudmundsson and Eliasson (1991).

DSC-measurements performed on several cereal starches show that they differ in their retrogradation tendency (Fig. 5). Most inclined for retrogradation is maize starch, whereas oat starch shows much less retrogradation. It is tempting to suggest that the differences illustrated in Fig. 5 are related to the amount of lipids present, or to their composition. Defatting of oat starch, which has a very high lipid content compared with other starches, results in higher retrogradation (Paton, 1987; Gudmundsson and Eliasson, 1989). Wheat starch, which shows rather low retrogradation, contains a large proportion of lysolecithin, whereas maize starch contains free fatty acids (Morrison, 1985).

When emulsifiers are added to starch ΔH_c decreases compared with the situation without an additive (Eliasson, 1983; Russell, 1983; Krog et al, 1989). The influence of polar lipids on the retrogradation of starch has been explained by the formation of the amylose-lipid complex. As described above, amylose-lipid complexes are present (the V-pattern) at increased levels (DSC-results) when emulsifiers are added. However, this complex-formation can not be the sole explanation. As it is amylopectin that retrogrades during the time period corresponding to the storage of bread, it is not evident why the amylose-lipid complex should have an influence. It might simply be that it does not. It has been observed that the addition of separately prepared amylose-lipid complexes to waxy maize starch, a starch essentially free of amylose, does not reduce or delay the retrogradation (Gudmundsson, 1992a). However, if the starch with added complexes was heated to above the transition temperature of the amylose-lipid complex before storage, the subsequent retrogradation decreased. The retrogradation of waxy maize starch in the presence of added surfactants (including MG) has been studied in the DSC, and the results in Fig. 6 were obtained. The added lipid, thus, decreases the retrogradation also of a waxy starch.

When triglycerides were added to a waxy starch at the level 1.4% (calculated on starch) no influence on retrogradation was observed (Eliasson and Ljunger, 1988). When increased levels are used, then a decrease in ΔHc has been measured (Silverio and Eliasson, 1994). This result is in accordance with firmness measurements (Rogers et al, 1988).

It has been observed that it is still possible to obtain an effect when adding surfactants to a system already containing shortenings (or vice versa) (Rogers et al, 1988). Emulsifiers and shortenings, thus, do not influence starch retrogradation in the same way. The effects of triglycerides could simply be to prevent the growth of amylopectin crystals by diluting them, whereas polar lipids might function through complex formation.

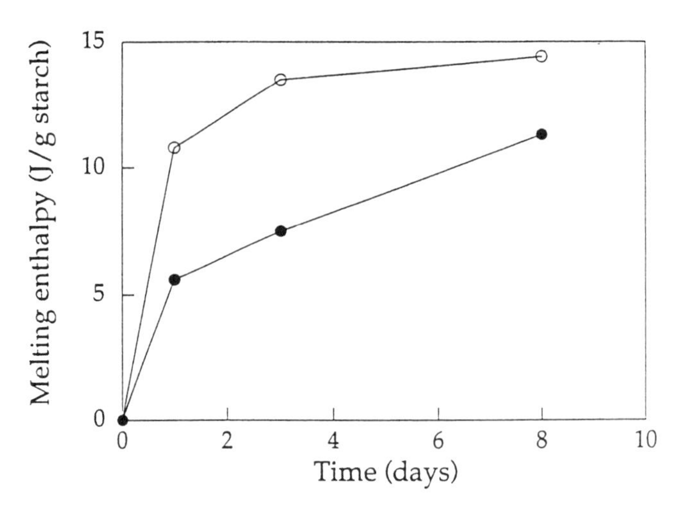

Figure 6. The retrogradation of a waxy starch in the presence of a surfactant (CTAB), O = waxy maize starch without added surfactant, ● = waxy maize starch with added CTAB. Data taken from Eliasson and Ljunger (1988).

Rheological Properties

In the section about gelatinization it was described how amylose leaching and starch granule swelling were affected by the presence of lipids. It is also true that breads with emulsifiers are softer than bread without. These results show that the rheological properties of a starch gel are greatly influenced by lipids. In this connection triglycerides and diacyl lipids (e.g. lecithin) seem to exert less influence than monoacyl lipids (Eliasson et al, 1988; Kim et al, 1992). This is not to say that triglycerides or diacyl lipids do not influence the rheological properties of a starch gel, only that the influence seems to be rather small when added at low levels (below five per cent). When triglycerides are added at levels used in high-fat bakery products they will certainly influence rheological properties. This becomes very evident in low-fat products, that frequently are found to be more firm than the full-fat counterpart.

A starch gel can be described as a two-phase system, where the starch granules (more or less fragmented) constitute the dispersed phase ("fillers") and where amylose/amylopectin leached from the granules constitute the continuous phase (Eliasson and Bohlin, 1982; Ring, 1985). The rheological properties of a starch gel depend on the phase volume of starch granules, on their deformability and on the amount of amylose/amylopectin leached (Ring, 1985; Eliasson, 1986b; Steeneken, 1989). Interactions between the two phases might also influence the behaviour of the gel (Eliasson, 1986b). Lipids, thus, influence the rheological properties of a starch gel by affecting all these parameters, either in combination or separately. Lipids will decrease the swelling of starch granules, and thus affect their phase volume as well as their deformability. Lipids will complex with amylose and thus influence the composition of the continuous phase. The amylose-lipid complex is insoluble, and can be expected to precipitate on available surfaces, like the starch granule surface, and then interactions between continuous and dispersed phases are affected.

The influence of the native lipids on, for example, wheat and maize starch gels can perhaps be understood from a comparison with potato starch which contain no lipid. This starch swells much more and gives a higher viscosity at an earlier temperature than wheat starch when compared in the Brabender Viscograph. During a holding period at about 95°C the potato starch paste is broken down much more than the wheat starch paste. A similar pattern is observed when the gelatinization is followed in fundamental rheological measurements (oscillatory measurements) (Eliasson, 1986c). The storage modulus (G') reached its highest value at temperatures around 75°C in potato starch, whereas wheat and maize starches had to be heated to 95°C. When lipids are added to potato starch the paste properties become more similar to those of wheat starch (Eliasson et al, 1981). One such property, that is immediately observed, is the appearance of the starch gel. Potato starch gels are transparent, but when lipids are added they become as opaque as the wheat starch gel. The texture will also change, from long in the potato starch gel to short in the potato starch gel with added lipids (or the wheat starch gel).

As added polar lipids delay the swelling of starch granules this can explain the lower viscosity at low temperatures. In addition, the decreased swelling will cause less sensitivity to shearing at elevated temperatures, and then cause increased viscosity. The effect of an addition of a lipid to a starch will then depend on the balance between swelling, break down and resistance against deformation.

The network structure, necessary for a gel to form, might be built up from the granules, or the granule fragments. However, the amylose-lipid complexes take part in network structures (Kim et al., 1992; Eliasson and Kim, 1995). The viscoelastic properties of starch gels with added MG prepared in the Brabender Viscograph, were studied in dynamic rheological measurements during heating. An example of the kind of result obtained is given in Fig. 7.

When a potato starch gel (or a waxy maize starch gel) is reheated after its preparation G' decreased monotonically, and the phase angle (δ) increased, indicating a loss of structure (Kim et al, 1992). After cooling the original values were restored. However, the changes were not large for the potato starch gel, δ increased from $\approx 39^o$ at room temperature to $\approx 43^o$ at 90°C. When a monoglyceride (monolaurin) was added the result in Fig. 7 was obtained. At room temperature the lipid-containing gel was different from the potato starch gel, G' had increased from $\approx$ 10 Pa to 184 Pa, and δ has decreased from $\approx 39^o$ to 6.4^o. The addition of monolaurin resulted in a more elastic (i.e. lower δ

values) and stiffer gel (higher G'). Heating changed this, and at 90°C the visco-elastic values were similar to the potato starch gel without added lipids.

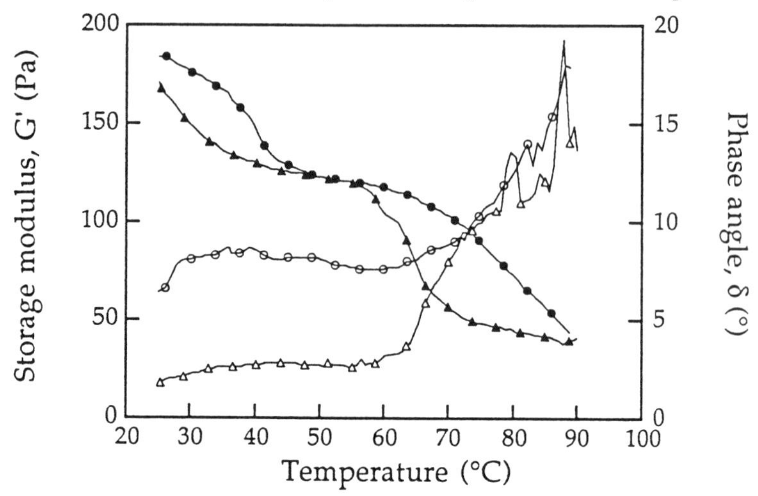

Figure 7. The changes in storage modulus (G') and phase angle (δ) during heating and cooling of a potato starch gel with monolaurin, ● = G' during heating, ▲ = G' during cooling, O = δ during heating, Δ = δ during cooling. Data adapted from Eliasson and Kim (1995).

The changes in G' as well as δ occurred mainly in the temperature range 70 - 90°C. These temperatures correspond to the melting temperature of the amylose-lipid complex with monolaurin as the ligand (Eliasson and Krog, 1985). When the sample in Fig. 7 was cooled the transitions were reversed, i.e. G' increased and δ decreased. This could be interpreted as formation of the amylose-lipid complex. There was also a transition in the heating curve at a lower temperature, at 40-50°C (Fig. 7). It is tempting to suggest that this is a transition related to amylopectin-lipid interactions, as this transition was present also for a waxy maize starch-monomyristin paste (Kim et al, 1992). An alternative interpretation could be that there is a temperature difference between formation of the complex and its crystallisation (cf. Biliaderis and Galloway (1989)).

The type of lipid present affects rheological properties. The G' increases and δ decreases with increasing chain length. When compared at the same chain length non-ionic ligands, like monoglycerides, give the strongest gel, whereas cationic ligands give the weakest (Eliasson and Kim, 1995).

Influence on Lipid Properties

So far the properties of the starch-lipid system, and how the addition of lipids influences starch properties, have been discussed. The complex-formation between amylose and lipids will also affect the lipid in different ways.

When amylose is present lipid monomers will be taken out from the system due to the precipitation of the complex. Thus, the equilibrium between monomers and liquid-crystalline phases will be shifted to the right:

$$\begin{matrix}\text{liquid} & & & & \text{amylose} \\ \text{crystalline} & \leftrightarrows & \text{monomer} & \rightarrow & \text{lipid} \\ \text{phase} & & & & \text{complex}\end{matrix}$$

This might influence what lipid phase can be formed. The amount of lipids available for other interactions decreases. Depending on the surrounding of the complex, polar lipids might prefer the complex or a lipid environment like lecithin (Eliasson, 1985c).

In the same way as enzymes do not as readily recognise amylose in the complex (see below) the lipid molecule will not be recognised. Moreover, unsaturated lipids in the complex are protected against oxidation (Acker, 1977). Complex formation can be a way to protect sensitive substances like aroma compounds or drugs (Wyler and Solms, 1982). During baking, and the subsequent staling of bread, it is very plausible that some of the aroma compounds are lost because of complexing with starch. The refreshening of stale bread, involving heating to melt crystallised amylopectin, might also release some of these aroma compounds.

MECHANISMS FOR THE STARCH-LIPID INTERACTION

So far in this chapter several references have been made to the amylose-lipid complex, but also other possibilities for starch-lipid interactions have been indicated, especially the amylopectin-lipid complex.

Formation of the Amylose-Lipid Complex

The structure of the amylose-lipid complex, illustrated in Fig. 1, has been determined from, among other techniques, X-ray diffraction studies (French and Murphy, 1977). The amylose molecule forms a helix around the hydrocarbon chain of the polar lipid. All three turns are not necessarily occupied, probably much less is required for the precipitation of the complex

(Carlson et al, 1979a). Usually each turn of the helix contains six glucose units, although seven and even eight units have been suggested for larger and more bulky ligands. The hydrocarbon chain in the complex is crystalline (Carlson et al., 1979a). For precipitation of the amylose-lipid complex a lipid:amylose ratio of 1:4 - 1:5 is frequently used (Riisom et al., 1984). However, not all of the added emulsifier will then complex, but part of it will be co-precipitated or adsorbed on the complex (Eliasson and Krog, 1985). In case of SDS it has been calculated that about 40-54% of the amylose are occupied, thus leaving 47- 60% of the space vacant (Yamamoto et al., 1983). Saturation molar ratios have been calculated for fatty acids and found to be 0.090 for capric acid and 0.054 for stearic acid (Raphaelides and Karkalas, 1988). This corresponds to 86% of the amylose molecule being occupied.

The structure of the amylose-lipid complex is similar to the amylose-iodine complex, and this is important to keep in mind. Iodine and lipids compete for the same location in the amylose molecule, and if iodine is used for quantitative determination of amylose, the starch has to be defatted. Otherwise the amylose complexed with lipids will not be analysed, and the total amount in the starch thus underestimated (Morrison and Laignelet, 1983).

For the influence of the amylose-lipid complex on the cereal product quality it is of course of interest to know how, when and where it is formed. Three different situations can be identified: complex formation in solution, complex formation in concentrated systems when heat is applied, and complex formation when heat and mechanical treatment is applied. The first situation corresponds to the laboratory preparation of complexes, and it is from such preparations that most of our knowledge about the complex stems. The second situation corresponds to different baking and cooking procedures, and the third situation corresponds to other ways of processing, e.g. extrusion cooking or drum drying.

In experimental situations the complex is easily formed: when a suitable lipid dispersion is added to an amylose solution the complex precipitates, and it can be collected by centrifugation for further studies. As this is a molecular interaction the possibilities for an interaction are better when both components are present as a molecular solution. This means that in case of lipids they should be present as micellar solutions or dispersed as liposomes, and for amylose treatment with strong alkali or DMSO (dimethyl sulphoxide) is used for dissolution. In cereal products it is usually not possible to affect the solubility of either component. The only exception is for added emulsifiers, where the addition of the emulsifier in the proper phase increases its efficiency. In this connection it is noteworthy that unsaturated monoglycerides, that are regarded as rather poor complexing agents, can be very effective if they are added as liposomes, instead of as the cubic phase (Riisom et al, 1984).

During many years it has been debated whether the amylose-lipid complex exists in the native starch granule or not. The difficulties to extract the lipids from starch have been taken as an argument for the native amylose-lipid complex, whereas the absence of a V-pattern for native starch has been taken as an argument against the existence of the native amylose-lipid complex. Recently, the use of ^{13}C CP/MAS NMR (^{13}C-cross-polarization/magic-angle spinning nuclear magnetic resonance) has solved the dispute, and it seems that the complex exists in the native starch granule (Morrison et al, 1993). During processing at room temperature, and when water is added, no further complex formation is to be expected, except perhaps for complex formation with amylose released from damaged starch.

During heating both lipids and amylose will be available to a greater extent, and then formation of new complexes occurs. An interesting point here is the competition for the lipids in such complicated systems as cereal products. In several investigations it has been shown that the lipid distribution in a dough changes during the baking process, and in the final product most of the lipids are bound to the starch (DeStefanis et al., 1977). If the lipids are of importance for the stability of the air/water interface during the oven spring they should be able to adsorb to the interface, and complexation with starch will certainly prevent them from that. Complexation with starch means that the effective lipid content decreases.

The processing of cereals affects the properties of amylose-lipid complexes formed (Stute and Konieczny-Janda, 1983; Björck et al, 1984). A few examples on how the thermal transition of the complex is affected are given in Table IV.

Evidently all processes influence the thermal properties of the amylose-lipid complex. Boiling, for example, increases the transition enthalpy, probably reflecting that more complexes have been formed. The V-pattern has been observed for a freshly pasted maize starch gel (Mestres et al, 1988). This V- pattern remained after acid hydrolysis of the starch gel. Drum-drying evidently has the most profound influence on the complexes, more perfect crystals seem to form, as judged from the sharpness, and the size of the endotherm (Table IV). Extrusion cooking, on the other hand, might result in a complete amorphous structure, i.e. no V-pattern, and no DSC-endotherm for the transition of the amylose-lipid complex (Mestres et al, 1988). The properties of the complex, obtained after processing, might be related to annealing during the process, i.e. keeping of the complex for a prolonged period of time at a temperature just below its melting temperature.

TABLE IV
Temperature at peak maximum (T), temperature range (ΔT) and enthalpy (ΔH) of the transition of the amylose-lipid complex after processing.

Wheat flour	T (°C)	ΔT (°C)	ΔH (J/g d.b.)
BOILING[a,b]			
Raw	91.9	14.7	1.1
Boiled	100.9	21.4	2.3
DRUM DRYING[b,c]			
Raw	91.9	14.7	1.1
Drum-dried	97.1	6.0	5.0
Drum-dried and reheated	93.6	24.0	2.4
Drum-dried and homogenized	96.1	19.0	2.3
Drum-dried with 2 % soya oil	92.7	9.7	4.3
Drum-dried with 1% linoleic acid	92.3	8.1	5.7
EXTRUSION[c]			
Raw	90.8	20.7	2.0
Extruded	87.6	15.2	1.8
with 2 % soya oil	85.6	12.4	0.44
with 1 % linoleic acid	84.1	10.5	0.3

[a]All DSC measurements performed with excess water.
[b]Data from Björk et al. (1984).
[c]Data from Schweizer et al. (1986).

When lipids are added before extrusion cooking, or other cereal processing methods, the outcome with regard to the amylose-lipid complex seems to be related to the conditions (Table IV). For extrusion cooking in the presence of monoglycerides the transition enthalpy was found to increase with increasing level of added MG (Galloway et al, 1989). The degradation of macromolecules in extrusion cooking is less extensive when lipids are present (Colonna and Mercier, 1983). This was attributed to the lipids acting as a lubricant during processing. The properties of the extruded material will also depend on the lipid content during extrusion cooking. For extruded wheat starch and wheat flour it has been observed that defatting increased expansion, whereas adding lipids decreased the expansion (Faubion and Hoseney, 1982). Removal of lipids also caused increased textural strength, whereas the addition of lipids decreased strength.

Some Properties of the Amylose-Lipid Complex

The amylose-lipid complex has the ability to crystallise, and in X-ray diffraction the V-pattern is obtained. The melting of the crystalline structure has been carefully investigated using the DSC (cf. (Eliasson, 1994)). At certain conditions, i.e. with respect to the lipid in the complex, and to the complex formation temperature, two thermal transitions are observed (Raphaelides and Karkalas, 1988; Biliaderis and Galloway, 1989). Two types of complexes thus seem to exist for some ligands. Of these, only one form gives rise to the V-pattern in the X-ray diffraction technique (Biliaderis and Galloway, 1989). The form melting at the lowest temperature might be transformed into the high-temperature melting form by annealing (Biliaderis and Galloway, 1989; Biliaderis and Seneviratne, 1990). For the crystalline regions a lamellar structure including chain folding has been proposed (Jane and Robyt, 1984). It might be speculated that the influence of the complex on starch properties will depend on its crystallinity (see below).

The transition temperature of the complex, observed in the DSC, depends on the type of lipid present, and it increases with increasing chain length and decreases with increasing degree of unsaturation of the ligand (Stute and Konieczny-Janda, 1983; Eliasson and Krog, 1985; Kowblansky, 1985; Raphaelides and Karkalas, 1988). The transition temperature at high water contents (i.e. more than 80% water) is around 90°C for myristic acid (90.3°C) and glycerol monomyristin (90.2°C), whereas it is around 100 C° for the corresponding stearic compounds (98.3 and 103.5°C, respectively) (Eliasson and Krog, 1985; Raphaelides and Karkalas, 1988). The influence of the polar head is related to the chain length, and for short chain lengths the non-ionic compounds have the highest transition temperature, whereas at longer chain length the ionic ligands might give higher transition temperature (Kowblansky, 1985). The transition temperature of the amylose-lipid complex is sensitive to the water content, and increases with decreasing water content (Eliasson, 1980; Biliaderis et al, 1986; Jovanovich et al, 1992). Thus, the transition of the complex is not likely to occur in most cereal products, except when short-chained ligands are present at high water contents.

When emulsifiers are added to delay the staling of bread, an increased enthalpy value is obtained for the transition of the amylose-lipid complex. This result indicates that more complexes are formed. The value does not change during storage (Mestres et al, 1988; Krog et al, 1989). As the enthalpy of the endotherm due to melting of retrograded amylopectin change during storage, these two endotherms are not related.

The amylose-lipid complexes form gels in the absence of starch, as has been shown for CTAB, monoglycerides, and fatty acids (Eliasson, 1988; Biliaderis and Galloway, 1989; Raphaelides, 1992). The mechanical spectra of gels formed by amylose-lipid complexes show a slight frequency dependence (Eliasson, 1988; Biliaderis and Galloway, 1989). Compared to an amylose control gel, addition of a fatty acid can increase or decrease G', depending on the concentration (Raphaelides, 1992). The strength of the amylose-lipid complex gel depends on the type of ligand, its concentration, and which form is present.

The enzymatic degradation by α-amylases is related to the type of lipid present in the complex, and the stability against hydrolysis parallels the thermal stability, i.e. it increases with the hydrocarbon chain length, and decreases with unsaturation (Eliasson and Krog, 1985). It has already been noted that the complexes are rather resistant against acid hydrolysis (Mestres et al, 1988). Also for the acid hydrolysis the type of complex present influences the hydrolysis rate (Biliaderis and Seneviratne, 1990).

Amylopectin-Lipid Complex

As already mentioned several experimental results can not be explained by the formation of the amylose-lipid complex. Most confusing is perhaps the explanation of how emulsifiers influence the staling of bread. Now and then it is said that the emulsifiers delay the staling of bread by preventing the crystallisation of amylose. It might be true that the crystallisation of amylose is prevented, but at the same time there appears to be no relation between crystallisation of amylose and staling of bread. The retrogradation of amylopectin can be delayed by adding emulsifiers, therefore the presence of an amylose-lipid complex is not necessary for obtaining that effect (Eliasson and Ljunger, 1988; Gudmundsson, 1992b).

Studies of the influence of lipids on the rheological properties of different starches show that waxy varieties are also affected (Evans, 1986; Eliasson et al, 1988). Waxy maize starch shows increased viscosity in the presence of SDS (Evans, 1986). When monomyristin was added to waxy maize starch at 95°C the gel that formed after cooling to room temperature gave higher G' and G'' values, and lower phase angle compared with the control without added MG (Kim et al, 1992). When the effects of CTAB, saturated monoglycerides, lecithin and soybean oil (all added at 2% on the starch) on the rheological behaviour of a cross-linked waxy maize starch were studied only CTAB was found to give a significant increase in G' (Eliasson et al, 1988).

In DSC-studies it was found that when the enthalpy of the transition of the amylose-lipid complex and the gelatinization enthalpy in the presence of lipids are summarised, the sum is smaller than the gelatinization enthalpy without added lipids. The difference could be interpreted as being due to the exothermic formation of an amylopectin-lipid complex (Evans, 1986; Eliasson et al, 1988).

Different studies of lipid-binding (equilibrium dialysis and surface tension measurements) show that lipids are "bound" to amylopectin or waxy starches (Hahn and Hood, 1987; Gudmundsson, 1992a). In equilibrium dialysis it was found that normal maize bound approximately seven times more lipid than did waxy maize (Hahn and Hood, 1987). Surface tension measurements gave a binding ratio of 17.5 of amylose versus amylopectin (Gudmundsson, 1992a). In the former case the binding of stearic acid was studied, and in the latter SDS.

The cited results are indirect evidence for the formation of an amylopectin-lipid complex. Direct evidence is more scarce. In contrast to the case of the amylose-lipid complex no precipitate is formed when lipid dispersions are added to amylopectin solutions. The formation of an insoluble complex has been described when amylopectin and monoglycerides were stirred at 60°C for long times (Batres and White, 1986). When waxy maize starch and lipids are heated in the DSC no endotherm is found, and the V-pattern is not obtained for the starch gel. Therefore, to obtain direct evidence seems to be difficult. Presumably it is the outer branches of the amylopectin molecule that are the sites for such complex formation to occur. If the minimum chain length for the formation of an amylose-lipid complex is 18 glucose units (Fig. 1), and the average degree of polymerisation for waxy cereal starches is 23-24 glucose units (Hizukuri, 1985), there is certainly a possibility for complexing to occur. For a melting endotherm to be detected in the DSC a rather high degree of cooperativity is required, and for the detection by X-ray diffraction a certain minimum crystal size is required. Probably, none of those conditions are fulfilled, except in certain situations (Slade and Levine, 1987; Gudmundsson and Eliasson, 1990; Huang and White, 1993).

Interactions without Complex-formation

Even if the existence of an amylopectin-lipid complex is accepted, such a complex can not explain all of the effects observed when starch and lipid are mixed. The effect of triglycerides on the retrogradation of amylopectin can hardly be related to the formation of a complex, as triglycerides have been shown not to complex with either amylose or with amylopectin. Instead a mechanism based on the disturbance of crystal growth might be considered. The presence of a fat simply prevents the crystals from growing into large structures that can form a three-dimensional network. If the formation of an amylose-network is essential for the product structure a large amount of fat may prevent such a network from forming.

Adsorption of lipids onto starch granules can certainly influence the rheological properties of a gel by changing the interaction between fillers and continuous phase. A lipid layer, surrounding the starch granule, will also change the water distribution into the starch granules (Larsson, 1983), something that could influence retrogradation.

As previous discussed, the amylose-lipid complex exists in different forms, and it might be speculated that the different forms have different effects. The differences in gel strength for potato starch gel prepared in the presence of ligands with a chain length of twelve, and different polar heads, might be related to different crystallinity of the complexes (Eliasson and Kim, 1995).

CONCLUSIONS

Starch-lipid interactions are the most abundant among carbohydrate-lipid interactions in cereal processing and cereal products. Such interactions affect the behaviour of starch as well as lipids, and will greatly affect product quality. Although the formation of the amylose-lipid complex can explain many of the effects observed this is not the only mechanism, and other possibilities are suggested.

Interactions between other polysaccharide present in cereals and lipids might be used to control rheology in the development of new products. Because of the complexity of cereal products there might be situations were there is a competition between available components for interacting with lipids. Such competition might influence the product quality.

REFERENCES

ACKER, L. 1977. Die Lipide der Stärken - ein Forchungsgebiet zwischen Kohlenhydraten und Lipiden. Fette-Seifen-Anstrichmittel 79: 1.

BATRES, L. R., and WHITE, P. J. 1986. Interaction of amylopectin with monoglycerides in model systems. J.A.O.C.S. 63: 1537-1540.

BILIADERIS, C. G., and GALLOWAY, G. 1989. Crystallization behaviour of amylose-V complexes: structure-property relationships. Carbohydr. Res. 189: 31-48.

BILIADERIS, C. G., PAGE, C. M., and MAURICE, T. J. 1986. Non-equilibrium melting of amylose-V complexes. Carbohydr. Polym. 6: 269-288.

BILIADERIS, C. G., and SENEVIRATNE, H. D. 1990. On the supermolecular structure and metastability of glycerol monostearate - amylose complex. Carbohydr. Polym. 13: 185-206.

BJÖRCK, I., ASP, N.-G., BIRKHED, D., ELIASSON, A.-C., SJOBERG, L.-B., and LUNDQUIST, I. 1984. Effects of processing on starch availability in vitro and in vivo. II. Drum-drying of wheat flour. J. Cereal Sci. 2: 165-178.

CARLSON, T., LARSSON, K., and MIEZIS, Y. 1978. Phase equilibria and structures in the aqueous system of wheat lipids. Cereal Chem. 55: 168-179.

CARLSON, T., LARSSON, K., and MIEZIS, Y. 1980. Physical structure and phase properties of aqueous systems of lipids from rye and triticale in relation to wheat lipids. J. Disp. Sci. Techn. 1: 197-208.

CARLSON, T. L.-G. 1981. Law and order in wheat flour dough. Colloidal aspects of the wheat flour dough and its lipid and protein constituents in aqueous media. Lund, Lund University.

CARLSON, T. L.-G., LARSSON, K., DINH-NGUYEN, N., and KROG, N. 1979a. A study of the amylose-monoglyceride complex by Raman spectroscopy. Starch/Stärke 31: 222-224.

CARLSON, T. L.-G., LARSSON, K., MIEZIS, Y., and POOVARODOM, S. 1979b. Phase equilibria in the aqueos system of wheat gluten lipids and in the aqueous salt system of wheat lipids. Cereal Chem. 56: 417-419.

CARLSSON, A., KARLSTRÖM, G., and LINDMAN, B. 1990. Thermal gelation of nonionic cellulose ethers and ionic surfactants in water. Colloids. Surf. 47: 147-165.

CARLSSON, A., KARLSTRÖM, G., LINDMAN, B., and STENBERG, O. 1988. Interaction between ethyl(hydroxyethyl)cellulose and sodium dodecyl sulphate in aqueous solution. Colloid Polym. Sci. 266: 1031-1036.

COLONNA, P., and MERCIER, C. 1983. Macromolecular modification of manioc starch components by extrusion-cooking with and without lipids. Carbohydr. Polym. 3: 87-108.

CROWE, J. H., CROWE, L. M., CARPENTER, J. F., RUDOLPH, A. S., AURELL-WISTROM, C., SPARGO, B. J., and AMCHARDOGUY, T. J. 1988. Interactions of sugars with membranes. Biochim. Biophys. Acta 947: 367-384.

DeSTEFANIS, V. A., PONTE, J. G., CHUNG, F. H., and RUZZA, N. A. 1977.
Binding of crumb softeners and dough strengtheners during breadmaking. Cereal Chem. 54: 13-24.

ELIASSON, A.-C. 1980. Effect of water content on the gelatinization of wheat starch. Starch/Stärke 32: 270-272.

ELIASSON, A.-C. 1983. Differential scanning calorimetry studies on wheat starch-gluten mixtures. II. Effect of gluten and sodium stearoyl lactylate on starch crystallization during ageing of wheat starch gels. J. Cereal Sci. 1: 207-213.

ELIASSON, A.-C. 1985a. Starch gelatinization in the presence of emulsifiers. A morphological study of wheat starch. Starch/Stärke 37:411-415.

ELIASSON, A.-C. 1985b. Retrogradation of starch as measured by differential scanning calorimetry. In: New Approaches to Research on Cereal Carbohydrates. R. D. Hill, and L. Munck, eds. Elsevier Science Publishers, Amsterdam, pages 93-98.

ELIASSON, A.-C. 1985c. Starch-lipid interactions studied by differential scanning calorimetry. Thermochim. Acta 95: 369-374.

ELIASSON, A.-C. 1986a. On the effects of surface active agents on the gelatinization of starch - a calorimetric investigation. Carbohydr. Polym. 6: 463-476.

ELIASSON, A.-C. 1986b. Viscoelastic behaviour during the gelatinization of starch. I. Comparison of wheat, maize, potato and waxy-barley starches. J. Text. Stud. 17: 253-265.

ELIASSON, A.-C. 1986c. Viscoelastic behaviour during the gelatinization of starch II. Effects of emulsifiers. J. Text. Stud. 17: 357-375.

ELIASSON, A.-C. 1988. On the thermal transitions of the amylose-cetyltrimethylammonium bromide complex. Carbohydr. Res. 172: 83-95.

ELIASSON, A.-C. 1989. Some physico-chemical properties of wheat starch. In: Wheat End-use Properties. Wheat and Flour Characterization for Specific End-use. H. Salovaara, ed. University of Helsinki, Helsinki, pages 355-364.

ELIASSON, A.-C. 1994 Interactions between Starch and Lipids Studied by DSC. Thermochim. Acta 246: 343-356.

ELIASSON, A.-C., and BOHLIN, L. 1982. Rheological properties of concentrated wheat starch gels. Starch/Stärke 34: 267-271.

ELIASSON, A.-C., FINSTAD, H., and LJUNGER, G. 1988. A study of starch-lipid interactions for some native and modified maize starches. Starch/Stärke 40: 95-100.

ELIASSON, A.-C., GUDMUNDSSON, M., and SVENSSON, G. 1995. Thermal behaviour of wheat starch in flour-relation to flour quality. Lebensm.-Wiss.u.- Technol. 28: 227-235.
ELIASSON, A.-C., and HEGG, P.-O. 1981. Thermal stability of gluten. Cereal Chem. 57: 436-437.
ELIASSON, A.-C., and KIM, H.-R. 1995. A dynamic rheological method to study the interaction between starch and lipids, J. Theol. 39: 1519-1534.
ELIASSON, A.-C., and KROG, N. 1985. Physical properties of amylose-monoglyceride complexes. J. Cereal Sci. 3: 239-248.
ELIASSON, A.-C., and LARSSON, K. 1993. Cereals in breadmaking: A molecular/colloidal approach. New York, Marcel Dekker.
ELIASSON, A.-C., LARSSON, K., and MIEZIS, Y. 1981. On the possibility of modifying the gelatinization properties of starch by lipid surface coating. Starch/Stärke 33: 231-235.
ELIASSON, A.-C., and LJUNGER, G. 1988. Interactions between amylopectin and lipid additives during retrogradation in a model system. J. Sci. Food Agric. 44: 353-361.
EVANS, I. D. 1986. An investigation of starch/surfactant interactions using viscometry and differential scanning calorimetry. Starch/Starke 38: 227-235.
FAUBION, J. M., and HOSENEY, R. C. 1982. High-temperature short-time extrusion cooking of wheat starch and flour. II. Effect of protein and lipid on extrudate properties. Cereal Chem. 59: 533-537.
FRENCH, A. D., and MURPHY, V. G. 1977. Computer modelling in the study of starch. Cereal Foods World 22: 61-70.
GALLOWAY, G. I., BILIADERIS, C. G., and STANLEY, D. W. 1989. Properties and structure of amylose-glyceryl monostearate complexes formed in solution or on extrusion of wheat flour. J. Food Sci. 54: 950-957.
GHIASI, K., HOSENEY, R. C., ZELESNAK, K., and ROGERS, D. E. 1984. Effect of waxy barley starch and reheating on firmness of bread crumb. Cereal Chem. 61: 281-285.
GODDARD, E. D. 1993a. Polymer-surfactant interaction Part I. Uncharged water-soluble polymers and charged surfactants. In: Interactions of Surfactants with Polymers and Proteins. E. Goddard, and K. P.Ananthapadmanabhan, eds. CRC Press, Boca Raton, pages 123-169.
GODDARD, E. D. 1993b. Applications of polymer-surfactant systems. In: Interactions of Surfactants with Polymers and Proteins. E. Goddard, and K. P. Ananthapadmanabhan, eds. CRC Press, Boca Raton, pages 395-414.
GOUGH, B. M., GREENWELL, P., and RUSSELL, P. L. 1985. On the interaction of sodium dodecyl sulphate with starch granules. In: New Approaches to Research on Cereal Carbohydrates. R. D. Hill, and L. Munck, eds. Elsevier, Amsterdam, pages 99-108.
GUDMUNDSSON, M. 1992a. Cereal starches Physicochemical properties and retrogradation. Lund, Lund University (thesis).

GUDMUNDSSON, M. 1992b. Effects of an added inclusion-amylose complex on the retrogradation of some starches and amylopectin. Carbohydr. Polym.17: 299-304.
GUDMUNDSSON, M., and ELIASSON, A.-C. 1989. Some physicochemical properties of oat starches extracted from varieties with different oil content. Acta. Agric. Scand. 39: 101-111.
GUDMUNDSSON, M., and ELIASSON, A.-C. 1990. Retrogradation of amylopectin and the effects of amylose and added surfactants/emulsifiers. Carbohydr. Polym. 13: 295-315.
GUDMUNDSSON, M., and ELIASSON, A.-C. 1991. Thermal and viscous properties of rye starch extracted from different varieties. Cereal Chem. 68: 172-177.
GUDMUNDSSON, M., and ELIASSON, A.-C. 1992. Comparison of thermal and viscoelastic properties of four waxy starches and the effect of added surfactant. Starch/Stärke 44: 379-385.
HAHN, D. E., and HOOD, L. F. 1987. Factors influencing corn starch-lipid complexing. Cereal Chem. 64: 81-85.
HIZIKURI, S. 1985. Relationship between the distribution of the chain length of amylopectin and the crystalline structure of starch granules. Carbohydr.Res. 141: 295-306.
HUANG, J. J., and WHITE, P. J. 1993. Waxy corn starch: monoglyceride interaction in a model system. Cereal Chem. 70: 42-47.
IZYDORCZYK, M., BILIADERIS, C. G., and BUSHUK, W. 1991. Physical properties of water-soluble pentosans from different wheat varieties. Cereal Chem. 68: 145-150.
IZYDORCZYK, M., and BILIADERIS, C. G. 1992. Influence of structure on the physicochemical properties of wheat arabinoxylan. Carbohydr. Polym. 17: 237-247.
JANE, J.-L., and ROBYT, J. F. 1984. Structure studies of amylose-V complexes and retrograded amylose by action of alpha amylases, and a new method for preparing amylodextrins. Carbohydr. Res. 132: 105.
JOHNSON, L. A. 1991. Corn: production, processing, and utilization. In: Handbook of Cereal Science and Technology. K. J. Lorenz, and K. Kulp, eds. Marcel Dekker, Inc., New York, pages 55-131.
JOVANOVICH, G., ZAMPONI, R. A., LUPANO, C. E., and ANON, M. C. 1992. Effect of water content on the formation and dissociation of the amylose-lipid complex in wheat flour. J. Agric. Food Chem. 40: 1789-1793.
JUNGE, R. C., HOSENEY, R. C., and VARRIANO-MARSTON, E. 1981. Effect of surfactants on air incorporation in dough and the crumb grain of bread. Cereal Chem. 58: 338-342.
KIM, H.-R., ELIASSON, A.-C., and LARSON, K. 1992. Dynamic rheological studies on an interaction between lipid and various native and hydroxypropyl potato starches. Carbohydr. Polym. 19: 211-218.

KOWBLANSKY, M. 1985. Calorimetric investigation of inclusion complexes of amylose with long-chain aliphatic compounds containing different functional groups. Macromolecules 18: 1776-1779.
KROG, N. 1971. Amylose complexing effect of food grade emulsifiers. Starch/Stärke 23: 206.
KROG, N., and NYBO-JENSEN, B. 1970. Interaction of monoglycerides in different physical states with amylose and their anti-firming effects in bread. J. Fd Technol. 5: 77-87.
KROG, N., OLESEN, S. K., TOERNAES, H., and JOENSSON, T. 1989. Retrogradation of the starch fraction in wheat bread. Cereal Foods World 34: 281-285.
KROG, N. J., RIISOM, T. H., and LARSSON, K. 1985. Applications in the food industry: I. In: Encyclopedia of Emulsion Technology Vol. 2 Applications. P.Becher, ed. Marcel Dekker, New York and Basel, pages 321-365.
KULP, K., and PONTE, J. G., Jr. 1981. Staling of white pan bread: fundamental causes. CRC Critical Reviews in Food Science and Nutrition 15: 1-48.
LARSSON, K. 1983. Physical state of lipids and their technical effects in baking. In: Lipids in Cereal Technology. P. J. Barnes, ed. Academic Press, London, pages 237-251.
LARSSON, K. 1986. Physical properties-structural and physical characteristics. In: The Lipid Handbook. F. D. Gunstone, J. L. Harwood, and F. B. Padley, eds. Chapman and Hall, London and New York, pages 321-384.
LEISSNER, O. 1986. A comparison of the effect of different polymorphic forms of lipids in breadmaking. Cereal Chem. 65: 202-207.
LINDMAN, B., and THALBERG, K. 1993. Polymer-surfactant interactions - recent developments. In: Interactions of surfactants with polymers and proteins. E. D. Goddard, and K. P. Ananthapadmanabhan, eds. CRC Press, Boca Raton, pages 203-294.
LORENZ, K. J. 1991. Rye. In: Handbook of Cereal Science and Technology. K.J. Lorenz, and K. Kulp, eds. Marcel Dekker, New York, pages 331-371.
MacRITCHIE, F., and GRAS, P. W. 1973. The role of flour lipids in baking. Cereal Chem. 50: 292-302.
MATTERN, P. J. 1991. Wheat. In: Handbook of Cereal Science and Technology. K. J. Lorenz, and K. Kulp, eds. Marcel Dekker, Inc, New York, pages 1-53.
MESTRES, C., COLONNA, P., and BULEON, A. 1988. Gelation and crystallization of maize starch after pasting, drum-drying or extrusion cooking. J. Cereal Sci. 7: 123-134.
MORRISON, W. R. 1978. Cereal lipids. In: Adv. Cereal Sci. Techn. Y. Pomeranz, eds. AACC, St. Paul, pages 221-348.

MORRISON, W. R. 1985. Lipids in cereals. In: New Approaches to Research on Cereal Carbohydrates. R. D. Hill, and L. Munck, eds. Elsevier, Amsterdam, pages 61-70.
MORRISON, W. R., and COVENTRY, A. M. 1985. Extraction of lipids from cereal starches with hot aqueous alcohols. Starch/Starke 37: 83-87.
MORRISON, W. R., and LAIGNELET, B. 1983. An improved colorimetric procedure for determining apparent and total amylose in cereal and other starches. J. Cereal Sci. 1: 9-20.
MORRISON, W. R., LAW, R. V., and SNAPE, C. E. 1993. Evidence for inclusion complexes of lipids with V-amylose in maize, rice and oat starches. J. Cereal Sci. 18: 107-109.
MUKERJEE, P., and MYSELS, K. J. 1971. Critical micelle concentrations of aqueous surfactant systems. Washington DC, Government Printing Office.
PATERNOTTE, T. A., ORSEL, R., and HAMER, R. J. 1994. Dynamic interfacial behaviour of gliadin-diacylgalactosylglycerol (MGDG) films: possible implications for gas-cell stability in wheat flour doughs. J. Cereal Sci. 19: 123- 129.
PATON, D. 1987. Differential scanning calorimetry of oat pastes. Cereal Chem. 64: 394-399.
POMERANZ, Y., SHOGREN, M. D., FINNEY, K. F., and BECHTEL, D.B. 1977. Fiber in breadmaking - effects on functional properties. Cereal Chem.54: 25-41.
RAJAPAKSA, D., ELIASSON, A.-C., and LARSSON, K. 1983. Bread baked from wheat/rice mixed flours using liquid-crystalline lipid phases in order to improve bread volume. J. Cereal Sci. 1: 53-61.
RAPHAELIDES, S., and KARKALAS, J. 1988. Thermal dissociation of amylose-fatty acid complexes. Carbohydr. Res. 172: 65-82.
RAPHAELIDES, S. N. 1992. Viscoelastic behaviour of amylose-fatty acid gels. J. Texture Stud. 23: 297-313.
RIISOM, T., KROG, N., and ERIKSEN, J. 1984. Amylose complexing capacities of cis- and trans- unsaturated monoglycerides in relation to their functionality in bread. J. Cereal Sci. 2: 105-118.
RING, S. G. 1985. Some studies on starch gelation. Starch/Starke 37: 80-83.
ROGERS, D. E., ZELEZNAK, K. J., LAI, C. S., and HOSENEY, R. C. 1988. Effect of native lipids, shortening, and bread moisture on bread firming. Cereal Chem. 65: 398-401.
RUSSELL, P. L. 1983. A kinetic study of bread staling by differential scanning calorimetry and compressibility measurements. The effect of added monoglycerides. J. Cereal Sci. 1: 297-303.
SCHWEIZER, T. F., REIMANN, S., SOLMS, J., ELIASSON, A.-C., and ASP, N.-G. 1986. Influence of drum-drying and twin-screw extrusion cooking on wheat carbohydrates, II, Effect of lipids on physical properties, degradation and complex formation of starch in wheat flour. J. Cereal Sci. 4: 249-260.

SILVERIO, J., and ELIASSON, A.-C. 1994. The effect of oils on the gelatinization and retrogradation of wheat starch. Quality Cereals in a Changing World, The Hague, (poster).
SLADE, L., and LEVINE, H. 1987. Recent advances in starch retrogradation. In: Industrial Polysaccharides The Impact of Biotechnology and Advanced Methodologies. S. S. Stivala, V. Crescenzi, and I. C. M. Dea, eds. Gordon and Breach Publishers, New York, pages 387-430.
SOULAKA, A. B., and MORRISON, W. R. 1985. The bread baking quality of six wheat starches differing in composition and physical properties. J. Sci.Food Agric. 36: 719-727.
STEENEKEN, P. A. M. 1989. Rheological properties of aqueous suspensions of swollen starch granules. Carbohydr. Polym. 11: 23-42.
STUTE, R., and KONIECZNY-JANDA, G. 1983. DSC-Untersuchungen an Stärken Teil II Untersuchungen an Stärke-Lipid-Komplexen. Starch/Stärke 35: 340-347.
SÖDERBERG, I., and LJUSBERG-WAHREN, H. 1990. Phase properties and structure of a monoglyceride-sucrose-water system. Chem. Phys. Lipids 55: 97-101.
THALBERG, K., LINDMAN, B., and KARLSTRÖM, G. 1990. Phase diagram of a system of cationic surfactant and anionic polyelectrolyte: tetradecyltri-methylammonium bromide - hyaluronan - water. J. Phys. Chem.94: 4289-4295.
Van GIJSSEL, J., and ELIASSON, A.-C. 1993. Effect of fat on rheological behaviour of high-fat doughs. Cereal Foods World 38: 610.
WYLER, R., and SOLMS, J. 1982. Über Stärke-Aromastoff-Komplexe. III. über die Stabilität von trockenen Aroma-Stärkekomplexen und anderen Pulveraromen. Lebensm. Wiss. u. technol. 15: 93-96.
YAMAMOTO, M., SANO, T., HARADA, S., and YASUNAGA, T. 1983. Cooperativity in the binding of sodium dodecyl sulfate to amylose. Bull.Chem. Soc. Jpn. 56: 2643-2646.
ZOBEL, H. F. 1973. A review of bread staling. Baker s Dig. 47(5): 52-61.

CHAPTER 4
PROTEIN-CARBOHYDRATE INTERACTIONS

K. R. Preston
Grain Research Laboratory, Canadian Grain Commission
1404-303 Main Street
Winnipeg, MB, Canada, R3C

INTRODUCTION

Proteins and carbohydrates are the major components of wheat and wheat flour. Protein content can vary from about 7% to about 15%. Depending on this value, carbohydrates can account for about 68-76% of total flour weight (14% moisture basis). The major carbohydrate present in wheat flour is starch (63-72% of flour weight). Other carbohydrates include pentosan (2-3%) and cellulose (about 0.6%).

The properties of these components have been extensively studied and their fundamental roles in determining processing quality have been well documented (see chapter 1). Interactions between carbohydrates and proteins can also strongly influence processing quality. Protein-starch interactions appear to be closely associated with endosperm hardness. During dough processing and baking, these interactions may influence dough rheological properties and gas retention properties. Evidence also suggests that protein-starch interactions can affect bread staling properties. Protein-pentosan interactions have also been documented (Udy, 1957; Jelaca and Hlynka, 1972; Yin and Walker, 1992). However, few studies are available showing the influence of these interactions on wheat and flour processing .

The emphasis of this chapter will be concerned with protein-starch interactions during processing.

ENDOSPERM HARDNESS

The nature of the starch-protein interface in wheat endosperm has a major influence on milling characteristics and the properties of the resulting flours (Sandstedt and Schroeder, 1960). In soft textured common wheats (Fig. la), breakage during milling occurs between that interface and through cell walls. In harder textured common wheats, breakage can occur along cell walls and across this interface (Fig. lb). Because of these differences, harder wheats produce larger particles of endosperm material when passed through corrugated (break) rolls. Harder wheats require more energy to mill but are easier to sift than softer wheats.

The higher levels of starch damage in harder wheats make them more suitable for the production of bread products since they have higher water absorption potential and, through the action of amylases, produce higher levels of sugars which can be utilized by yeast during fermentation and proofing. Softer wheats are preferred for products such as cookies and cakes where lower water absorption is desirable. Durum wheats have extremely hard kernel texture. During milling, very large endosperm particles \ (semolina) can be produced that are most suitable for pasta production.

Microscopic studies of transverse sections of endosperm and flour particles by Hess (1954), Sandstedt and Schroeder (1960) and other workers (Aranyi and Hawrylewicz, 1969; Simmonds, 1972: Hoseney and Sieb, 1973; Crozet and Guilbot, 1979) indicated stronger adhesion between starch granules and endosperm protein in harder wheats than in softer wheats. This difference in adhesion was suggested as the primary cause of differences in wheat hardness. Supporting evidence for this hypothesis was provided by the work of Barlow and co-workers (1973).

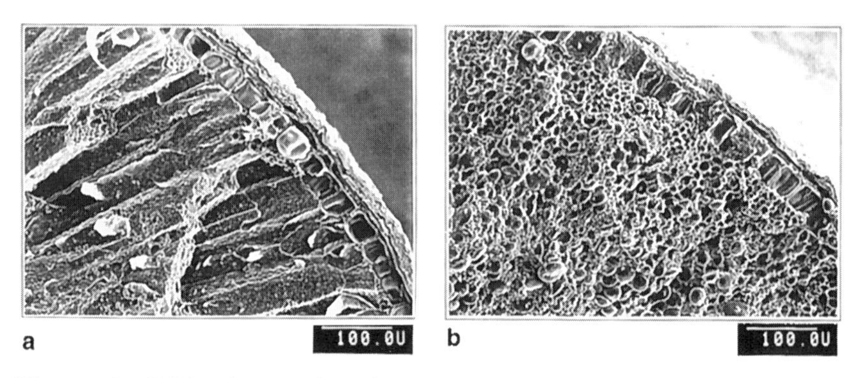

Figure 1. SEM micrographs of sections of hard (a) and soft (b) wheat endosperm (unpublished data).

Micropenetrometer testing of starch and storage protein separated by air classification and further purified by solvent suspension in mixtures of chloroform and benzene indicated little difference in hardness between starch or between protein from hard and soft varieties.

Fluorescent antibody staining and solvent extraction followed by electrophoresis were used by Barlow et al (1973) to show the presence of a layer of water soluble proteins surrounding isolated starch granules from both soft and hard wheats. Further studies (Simmonds et al, 1973) showed that the amount of water soluble protein associated with starch granules was dependent on wheat hardness, suggesting that these proteins may be important in determining this property. However, no specific proteins could be identified electrophoretically that were associated with these differences. The indistinguishability of gliadin - electrophorograms from hard and soft near-isogenic wheat lines used in the same study also provided evidence that the nature of wheat storage proteins was not a major determinant of wheat hardness.

In a more recent study, Greenwell and Schofield (1986) identified a lower (15K) molecular weight protein from water washed starch granules by high resolution SDS-polyacrylamide gel electrophoresis that gave a prominent band with soft wheats but only a faint band with harder wheats. It was also shown that the gene coding for this 15K band was on the same chromosome (5D) as the major gene identified previously by Law et al (1978), which controls endosperm texture. The authors used this evidence to suggest that the presence of this 15K protein, named friabilin, plays a major role in conferring endosperm softness on wheats. Furthermore, they suggested that since this protein is associated with the surface of starch granules in soft wheats, it may have properties that reduce the adhesion between the starch granule and the protein matrix of the endosperm. The biochemical properties of this protein have been studied recently (Sulaiman et al, 1993). Friabilin-like proteins have also been identified in other Triticeae species (Morrison et al, 1992, see also chapter VI).

Malouf and co-workers (Malouf and Hoseney, 1992; Malouf et al, 1992) fractionated hard and soft wheats into starch, gluten and water solubles. Measurement of the tensile strength of tablets produced from reconstitution experiments showed that this property was determined primarily by the starch fraction. Removal of the surface protein containing friabilin from soft wheat starch granules gave tablets with high tensile strength, similar to that obtained with hard wheat starch granules. These results support the hypothesis that friabilin plays a major role in determining wheat hardness.

Glenn and Saunders (1990) also reported that friabilin was evident only in soft wheats. However, texturally hard grains of genetically soft wheat contained this 15K protein. Furthermore, differences in the intensity of this protein among soft wheat varieties did not appear to be related to NIR hardness scores. These authors suggested that other factors, such as protein matrix structure, are also important in determining endosperm hardness. These findings are supported by work showing that cold steeping of hard wheat prior to tempering and milling can result in a reduction in starch damage to levels similar to values associated with soft wheats even though flow properties during milling are still similar to the untreated wheat (Preston et al, 1987).

DOUGH AND BREAD PROCESSING

Microscopy

Interactions between protein and starch occur during all stages of processing. Baker (1941) studied the formation and disruption of dough films during processing by light microscopy. His results indicated that the protein film formed during mixing and fermentation was closely associated with starch granules. During baking, this film expanded and was pulled out into gelatinous strings to which starch still adhered. Some of this adhesion between the gluten and the starch was reduced during this process. Light microscopic studies by Sandstedt et al (1954) confirmed these results. They noted that the adhesion between protein and starch was dependent upon water concentration. At low water concentrations, there appeared to be a strong bond between protein and starch whereas at excess water concentrations, these interactions were weak and the starch could easily be washed from the protein. This factor has been suggested as the major reason for the stronger apparent interactions between protein and starch which were observed microscopically in noodle (Moss et al, 1987) and steamed bread doughs (Huang and Moss, 1991) compared to bread doughs. Noodle doughs are generally mixed at 31-32% absorption, steamed bread doughs at 45-55% and bread doughs above 55% .

The changes occurring in the structure of doughs during processing have also been followed by various electron microscopy techniques. Evans et al (1981) reported that scanning electron micrographs of bread doughs during processing indicated that protein-starch interactions increased during mixing and decreased upon over mixing. They suggested that at optimum development, starch granules act as anchor points and facilitate formation of gluten fibrils upon stretching. Similar conclusions were reached by Cumming and Tung (1977).

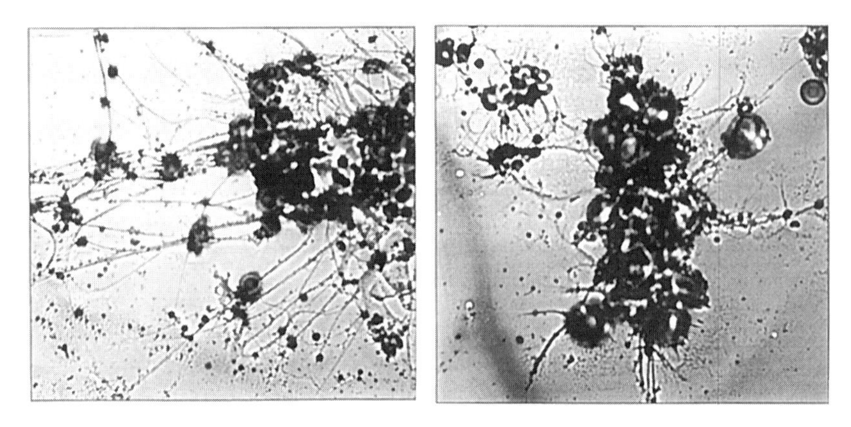

Figure 2. Micrograph of fibrils formed by wetting wheat endosperm particles (unpublished data).

Fretzdorff et al (1982) used freeze-fracture techniques to obtain scanning electron micrographs of bread doughs during processing and after baking. Their results suggested that protein-starch interactions intensified after fermentation and that after baking, protein and starch were tightly linked. Scanning electron microscopy studies by Pomeranz et al (1984) indicated that swollen starch, mainly from large starch granules, interacted most strongly with protein in bread. A number of other studies using electron microscopy have also demonstrated protein-starch interactions during bread processing (Fleming and Sosulski, 1978; Pomeranz and Meyer, 1984; Bechtel, 1985; Bechtel et al, 1978; Gan et al, 1989).

A number of ingredients, in addition to water, also appear to affect protein-starch interactions. Moss (1974, 1975) studied the effects of cysteine, potassium bromate and ascorbic acid on the microstructure of mixed, sheeted and proofed doughs and bread. Their results suggested that the film of gluten formed when doughs were over-mixed or mixed with excess cysteine resulted in stronger protein-starch interactions, resulting in a very cohesive and extensible dough. Moss et al (1987) showed that increased protein content appeared to strengthen protein- starch interactions while high pH reduced these interactions during noodle processing. Addition of gluten also appeared to strengthen protein-starch interactions during whole wheat processing (Gan et al, 1989). Addition of non-gluten proteins such as soy, sunflower, fababean and field pea had the opposite effect (Fleming and Sosulski, 1978). Cumming and Tung (1975) showed that removal of lipid increased protein-starch interactions.

Unfortunately, many of these studies must be viewed with some caution. Descriptions of the ultrastructural properties of the protein structure have varied widely, as pointed out by Freeman and Sheldon (1991). These differences have primarily been attributed to differences in sample preparation procedure which may mask the true nature of the protein structure (Varriano-Marston, 1977; Chabot et al, 1979, Freeman and Sheldon, 1991).

The initiation of protein structure during processing is probably related to the formation of protein fibrils which form extremely rapidly when flour particles are hydrated (Seckinger and Wolf, 1970; Bernardin and Kasarda, 1973ab; Amend and Belitz, 1991). The fibrils are pulled from endosperm particles by the surrounding flowing solution and exhibit both viscous flow and elastic deformation (Bernardin and Kasarda,1973a). In the same study, light and electron microscopy results showed that the protein fibrils adhered to starch granules. A photomicrograph of fibrils from wetted wheat flour, showing adhesion to starch granules, is shown in Fig. 2.

Microscopic and surface pressure studies indicate that surface forces, creating an interfacial protein film at the air-water interface, are responsible for this fibril formation (Amend and Belitz, 1989, Amend et al, 1990, Evers et al, 1990). This view is consistent with surface tension studies by Lundh et al (1988).

Recent studies by Amend and Belitz (1991) showed that the adhesion between protein fibrils and starch which occurred upon wetting of flour particles could be easily disrupted by gently stretching the fibrils which adhered to the glass cover slide. These results are consistent with studies cited earlier in this section indicating that high water concentrations reduce protein-starch interactions in doughs.

Wheat starch has been shown to possess properties that make it superior to other types of starch (rice, corn, waxy corn and potato) when mixed with gluten to produce bread (Harris, 1942; Sandstedt, 1961). Starches from different wheat varieties were also shown to affect bread quality produced from starch-gluten blends (Harris, 1942; Harris and Sibbitt, 1942; Soulaka and Morrison, 1985). Medcalf (1968) demonstrated that starch properties could also affect dough mixing properties. Starch-gluten mixograph curves revealed that wheat starch gave much stronger dough properties than corn starch. Differences in mixing properties were also found among wheat starches in starch- gluten blends. It was suggested that these differences may be related to starch properties which affect starch-protein interactions. Hoseney et al (1971) also showed differences in mixing requirements of bread doughs from reconstituted flours prepared with starches from different cereals.

Dynamic shear measurements by Cumming and Tung (1977) indicated that addition of increasing levels of starch to gluten did not change the character of the viscoelastic response but did cause an increase in the magnitude of the response. They concluded that this increase was primarily because of increased competition for available water. Lindahl and Eliasson (1986) studied interactions between gelatinized starch and wheat proteins by oscillatory rheological measurements. They found that the storage modulus, G', and the phase angle were sensitive to the composition of the starch gel. In the presence of gluten, G' increased for wheat and rye starches, decreased for corn starch and showed little change with barley, triticale and potato starches. A water soluble fraction containing lower MW protein (<65,000) gave similar effects. They attributed these changes to differences in the surface properties of the gelatinized starches which affect the interaction between the starch and protein.

Dreese et al (1988) measured the changes in dynamic rheological properties of flour-water doughs during heating. G' values decreased slowly between 25 and 50°C and then began to increase rapidly above 55°C, reaching a maximum value at about 70°C. Tangent values showed the opposite effect. Heating and cooling showed that irreversible changes were caused by heating above 55°C. Addition of gelatinized starch to starch-gluten blends also resulted in increased G' and decreased tangent values, suggesting that the irreversible changes in dough rheology due to heating could be attributed to this starch gelatinization. Decreasing dough moisture content increased G' values but had little effect upon tangent values. Thus starch gelatinization appears to affect doughs in other ways besides water absorption. The authors suggested that increased hydrogen bonding between gelatinized starch and gluten was primarily responsible for these changes.

He and Hoseney (1991) showed that a poor baking quality wheat flour had less oven spring and lost CO_2 at a much lower temperature (33°C) than a good baking quality wheat flour (72°C) when baked in an electrical heat resistance oven. Gluten-water doughs from the good variety showed higher G', lower tangent and lower protein solubility than gluten-water doughs from the poor variety over a broad temperature range (25-100°C), suggesting that stronger protein-protein interactions occurred in the former. In contrast, non-yeasted doughs from the poor quality variety showed higher G' and lower tangent values compared to values obtained with the better variety when heated to temperatures below 65°C. These results suggested that the poor quality flour had more effective cross-links than the good quality flour dough, presumably because of stronger starch- protein interactions. Recent studies by Petrofsky and Hoseney (1995) support this supposition. These workers showed that, in addition to gluten source, starch source had a significant effect heat starch doughs had higher moduli compared to hard wheat starch (Fig 3 and 4) when mixed with a commercial gluten.

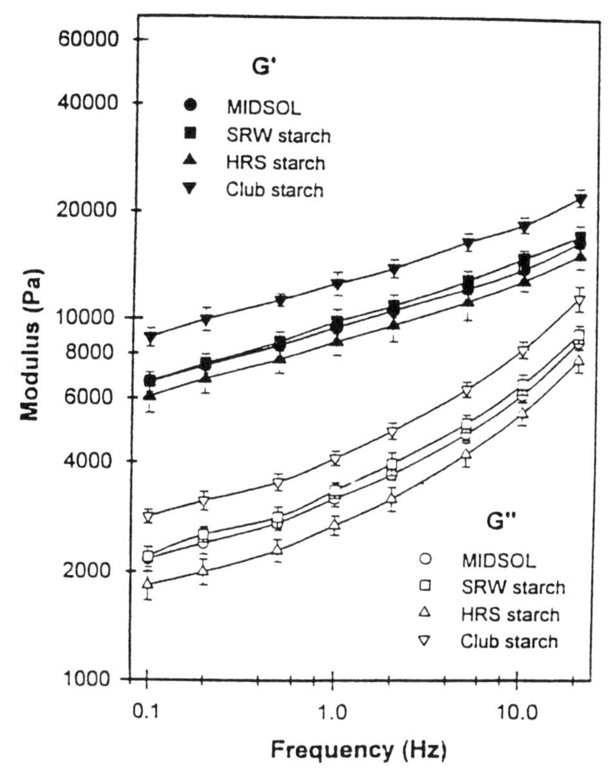

Figure 3. Storage modulus (G') and loss modulus (G") versus frequency for commercial gluten plus starches isolated from soft red winter, hard red spring and club wheat flours. The control (MIDSOL) is commercial gluten plus commercial starch. Reprinted with permission (Petrofsky and Hoseney, 1995).

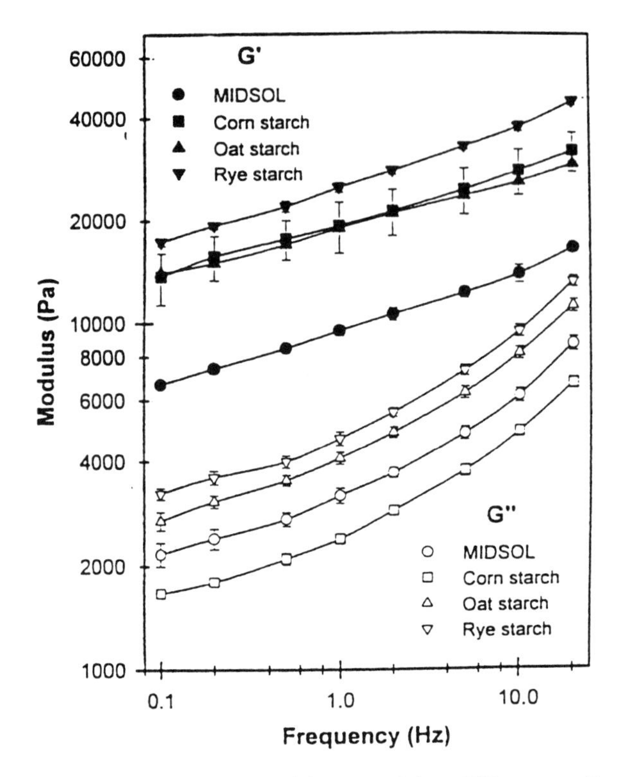

Figure 4. Storage modulus (G') and loss modulus (G") versus frequency for commercial gluten plus starches isolated from corn, oats and rye. The control (MIDSOL) is commercial gluten plus commercial starch. Reprinted with permission (Petrofsky and Hoseney. 1995).

STALING

Staling of bread has been shown to be primarily due to retrogradation of gelatinized starch, particularly amylopectin (see review by Kulp and Ponte, 1981). Erlander and Erlander (1969) showed that this process could be inhibited by complex formation of the starch polymers with lipids and proteins. It was postulated that proteins interact with the C-2 and C-3 hydroxyls of glucose units through H-bonding and prevent intermeshing of amylose and amylopectin helices. Support for these conclusions was obtained by Eliasson (1983) by differential scanning calorimetry (DSC). Using the Avrami equation to analyse DSC curves, addition of gluten to gelatinized starch resulted in a decrease in starch crystallization during ageing at 21 °C.

Recently, Hoseney and co-workers (Martin et al, 1991; Martin and Hoseney, 1991) have proposed that starch retrogradation may not be the major cause of bread firming during storage. On the basis of studies of firming rates of bread baked in an electrical heat resistance oven, it was postulated that bread firming was the result of increasing cross-linkage (H-bonding) between the protein matrix and the discontinuous remnants of starch granules during storage.

BINDING STUDIES

As discussed at the beginning of this chapter, wheat starch granules have a layer of lower MW "water-soluble" protein which is difficult to remove (Barlow et al, 1973; Simmonds et al, 1973; Greenwell and Schofield, 1986, Seguchi, 1986). Morrison and Scott (1986) have reported bound protein values in the range of 0.25-0.90 m^2/g of wheat starch. In addition, starch granules also appear to bind "gliadins" (Eliasson et al, 1981, Skerritt et al, 1990). Eliasson et al (1981) found variation in the amount of protein and lipid bound by starch granules from different varieties of wheat. They also suggested that the presence of these components could impart both hydrophobic and hydrophilic properties to the starch granule surface and suggested that variation in these components could affect the properties of starch, particularly gelatinization temperature.

In a more recent study, Eliasson and Tjerneld (1990), studied the absorption of proteins by starch granules in aqueous medium. They found that wheat starch absorbed more higher MW wheat protein (about 10 mg/m^2 starch) than lower MW wheat protein or bovine serum albumin. Absorption was also dependent on pH and ionic strength. They also found that potato starch absorbed much more protein than wheat or maize starch. Dahle (1971) showed that gelatinized starch also absorbs wheat proteins with optimum absorption occurring at lower pH. Heat denaturation of the protein component prevented absorption. More basic studies of these types are required to better understand protein-starch interactions and their relationship to processing properties and product quality.

REFERENCES

AMEND, T. and BELITZ, H. 1989. Microscopical studies of water/flour systems. Z Lebensm Unters Forsch 189:103-109.

AMEND, T., BELITZ, H. and KURTHEN, C. 1990. Electron microscopic studies on protein films from wheat and other sources at the air/water interface. Z Lebensm Unters Forsch 190:217-222.

AMEND. T. and BELITZ, H. 1991. Microstructural studies of gluten and a hypothesis on dough formation. Food Structure 10:277-288.

ARANYI, C. A. and HAWRYLEWICZ, E. J. 1969. Application of scanning electron microscopy to cereal specimens. Cereal Sci. Today 14:230-233,254.

BAKER, J. C. 1941. The structure of the gas cell in bread dough.Cereal Chem 18:34-41

BARLOW, K. K., BUTTROSE, M. S., SIMMONDS, D. H. and VESK, M. 1973. The nature of the starch-protein interface in wheat endosperm. Cereal Chem. 50:443-454.

BECHTEL, D. B. 1985. The microstructure of wheat: Its development and conversion into bread. Food Microstructure 4:125-133.

BECHTEL, D. B., POMERANZ, Y. and DE FRANCISCO, A. 1978. Breadmaking studied by light and transmission electron microscopy. Cereal Chem. 55:392-401.

BERNARDIN, J. E. and KASARDA, D. D. 1973a. Hydrated protein fibrils from wheat endosperm. Cereal Chem. 50:529-536.

BERNARDIN, J. E. and KASARDA, D. D. 1973b. The microstructure of wheat protein fibrils. Cereal Chem. 50:735-745.

CHABOT, J. F., HOOD, L. F. and LIBOFF, M. 1979. Effect of scanning electron microscopy preparation methods on the ultrastructure of white bread. Cereal Chem. 56: 462-464.

CROZET, N. and GUILBOT, A. 1979. Evolution des constituants proteiques et lipoproteiques de l'albumin de ble en cours de maturation. Ann. Technol. Agric. 28:211-222.

CUMMING, D. B. and TUNG, M. A. 1975. The ultrastructure of commercial wheat gluten. J. Inst. Can. Sci. Technol. Aliment. 8:67-73.

CUMMING, D. B. and TUNG, M. A. 1977. Modification of the ultrastructure and rheology of rehydrated commercial wheat gluten. Can. Inst. Food Sci.Technol. J. 10:109-119.

DAHLE, L. K. 1971. Wheat protein-starch interactions. I. Some starch-binding effects of wheat-flour proteins. Cereal Chem. 48:706-714.

DREESE, P. C., FAUBION, J. M., and HOSENEY, R. C. 1988. Dynamic rheological properties of flour, gluten, and gluten-starch doughs. I. Temperature-dependent changes during heating. Cereal Chem. 65:348-353.

ELIASSON, A.-C. 1983. Differential scanning calorimetry studies on wheat starch-gluten mixture. J. Cereal Sci. 1:207-213.

ELIASSON, A.-C., CARLSON, T. L.-G., LARSSON, K. and MEIZIS, Y. 1981. Some effects of starch lipids on the thermal and rheological properties of wheat starch. Starch 33:130-134.
ELIASSON, A.-C. and TJERNELD, E. 1990. Absorption of wheat proteins on wheat granules. Cereal Chem. 67:366-372.
ERLANDER, S. R. and ERLANDER, L. G. 1969. Explanation of ionic sequences in various phenomena. X. Protein-carbohydrate interactions and the mechanism for the staling of bread. Die Starke 21:305-315.
EVANS, L. G., PEARSON, A. M. and HOOPER, G. R. 1981. Scanning electron microscopy of flour-water doughs treated with oxidizing and reducing agents. Scanning Electron Microscopy P3:583-592.
EVERS, A. D., KERR, H. R. and CASTLE, J. 1990. The significance of fibrils produced by hydration of wheat proteins. J. Cereal Sci. 12:207-221.
FLEMING, S. E. and SOSULSKI, F. W. 1978. Microscopic evaluation of bread fortified with concentrated plant proteins. Cereal Chem. 55:373-382.
FREEMAN, T. P. and SHELDON, D. R. 1991. Microstructure of wheat starch: From kernel to bread. Food Technol. 45(3):162-168.
FRETZDORFF, B. BECHTEL, D. B. and POMERANZ, Y. 1982. Freeze-fracture ultrastructure of wheat flour ingredients, dough and bread. Cereal Chem. 59:113-120.
GAN, Z., ELLIS, P. R., VAUGHAN, J. G. and GALLIARD, T. 1989. Some effects of non-endosperm components of wheat and of added gluten on wholemeal bread structure. J. Cereal Sci. 10:81-91.
GLENN, G. M. and SAUNDERS, R. M. 1990. Physical and structural properties of wheat endosperm associated with grain texture. Cereal Chem. 67:176-182.
GREENWELL, P. and SCHOFIELD, J. D. 1986. A starch granule protein associated with endosperm softness in wheat. Cereal Chem. 63:379-380.
HARRIS, R. H. 1942. The baking quality of gluten and starch prepared from different wheat varieties. Baker's Digest 16(10):217-222,230,234.
HARRIS, R. H. and SIBBITT, L. D. 1942. The comparative baking qualities of hard red spring wheat starches and glutens as prepared by the gluten-starch blend baking methods. Cereal Chem.19:763-772.
HE, H. and HOSENEY, R. C. 1991. Differences in gas retention, protein solubility, and rheological properties between flours of different baking quality. Cereal Chem. 68:526-530.
HESS, K. 1954. Protein, gluten and lipid in wheat grain and flour Kolloid Z. 136:84-99.
HOSENEY, R. C., FINNEY, K. F., POMERANZ, Y., and SHOGREN, M. D. 1971. Functional (breadmaking) and biochemical properties of wheat flour components. VIII. Starch. Cereal Chem. 48:191-201.
HOSENEY, R. C. and SIEB, P. A. 1973. Structural differences in hard and soft wheat. Baker's Dig. 47(6): 26-28,56.

HUANG, S. and MOSS, R. 1991. Light microscopy observations on the mechanism of dough development in Chinese steamed bread production. Food Microstructure 10:289-293.
JELACA, D. L. and HLYNKA, I. 1972. Effect of wheat-flour pentosans in dough, gluten, and bread. Cereal Chem. 49:489-495.
KULP, K. and PONTE, JR., J. G. 1981. Staling of white pan bread. CRC Critical Reviews in Food Science and Nutrition 15(1):1-48.
LAW, C. N., YOUNG, C. F., BROWN, J. W. S., SNAPE, J. W. and WORLAND, A. J. 1978. The study of grain protein control in wheat using whole chromosome substitution lines. Page 483 in: Seed Protein Improvement by Nuclear Techniques. Int. Atomic Energy Agency, Vienna.
LINDAHL, L. and ELIASSON, A.-C. 1986. Effects of wheat proteins on the viscoelastic properties of starch gels. J. Sci. Food Agric. 37:1125-1132.
LUNDH, G., ELIASSON, A.-C. and LARSSON, K. 1988. Cross-linking of wheat storage protein monolayers by compression/expansion cycles at the air/water interface. J. Cereal Sci. 7:1-9.
MALOUF, R. B. and HOSENEY, R. C. 1992. Wheat hardness. I. A method to measure endosperm tensile strength using tablets made from wheat flour. Cereal Chem. 69:164-168.
MALOUF, R. B., LIN, W. D. A. and HOSENEY, R. C. 1992. Wheat hardness. II. Effect of starch granule protein on endosperm tensile strength. Cereal Chem. 69:169-173.
MARTIN, M. L. and HOSENEY, R. C. 1991. A mechanism of bread firming. II. Role of starch hydrolysing enzymes. Cereal Chem.68:503-507.
MARTIN, M. L., ZELEZNEK, K. J. and HOSENEY, R. C. 1991. A mechanism of bread firming. I. Role of starch swelling. Cereal Chem 68:498-503.
MEDCALF, D. G. 1968. Wheat starch properties and their effect on bread baking quality. Baker's Digest 42(4):48-50,52,65.
MORRISON, W. R., GREENWELL, P., LAW, C. N. and SULAIMAN, B. D. 1992. Occurrence of friabilin, a low molecular weight protein associated with grain softness, on starch granules isolated from some wheats and related species. J. Cereal Sci. 15:143-149.
MORRISON, W. R. and SCOTT, D. C. 1986. Measurements of the dimensions of wheat starch granules using a Coulter Counter with 100-channel analyzer. J. Cereal Sci. 4:13-21.
MOSS, R. 1974. Dough microstructure as affected by cysteine, potassium bromate, and ascorbic acid. Cereal Sci. Today 19:557- 561.
MOSS, R. 1975. Bread microstructure as affected by cysteine, potassium bromate, and ascorbic acid. Cereal Foods World 20:289- 292.
MOSS, R., GORE, P. J. and MURRAY, I. C. 1987. The influence of ingredients and processing variables on the quality and microstructure of hokkien, Cantonese and instant noodles. Food Microstructure 6:63-74.

PETROFSKY, K. E. and HOSENEY, R. C. 1995. Rheological properties of dough made with starch and gluten from several cereal sources. Cereal Chem. 72: 53-58 (1995).
POMERANZ, Y. and MEYER, D. 1984. Light and scanning microscopy of wheat- and rye-bread crumb. Interpretation of specimens prepared by various methods. Food Microstructure 3:159-164.
POMERANZ, Y., MEYER, D. and SEIBEL, W. 1984. Wheat, wheat-rye, and rye dough and bread studied by scanning electron microscopy. Cereal Chem. 61:53-59.
PRESTON, K. R., KILBORN, R. H. and DEXTER, J. E. 1987. Effects of starch damage and water absorption on the alveograph properties of Canadian hard red spring wheat. Can. Inst. Food Sci. Technol.J. 20:75-80.
SANDSTEDT, R. M. 1961. The function of starch in the baking of bread. Baker's Digest 25(3):36-42,44.
SANDSTEDT, R. M., SCHAUMBURG, L. and FLEMING, J. 1954. The microstructure of bread and dough. Cereal Chem. 31:43-49.
SANDSTEDT, R. M. and SCHROEDER, H. 1960. A photomicrographic study of mechanically damaged starch. Food Technol. 14:257-265.
SECKINGER, H. L. and WOLF, M. J. 1970. Electron microscopy of endosperm protein from hard and soft wheat. Cereal Chem. 47:236- 243.
SEGUCHI, M. 1986. Dye binding to the surface of wheat starch granules. Cereal Chem. 518-520.
SIMMONDS, D. H. 1972. The ultrastructure of the mature wheat endosperm. Cereal Chem. 49:212-222.
SIMMONDS, D. H., BARLOW, K. K. and WRIGLEY, C. W. 1973. The biochemical basis of grain hardness in wheat. Cereal Chem. 50:553-562.
SKERRITT, J. H., FREND, A. J., ROBSON, L. G. and GREENWELL, P. Immunological homologies between wheat gluten and starch granule proteins. J. Cereal Sci. 12:123-136.
SOULAKA, A. B. and MORRISON, W. R. 1985. The breadmaking quality of six wheat starches differing in composition and physical properties. J. Sci. Food Agric. 36:719-727.
SULAIMAN, B. D., BRENNAN, C. S., SCHOFIELD, J. D. and VAUGHAN, J.G. 1993. Some biochemical properties of friabilin and polyclonal antibody production. Aspects Applied Biology 36:61-68.
UDY, D. C. 1957. Interactions between proteins and Polysaccharides of wheat flour. Cereal Chem. 34:37-46.
VARRIANO-MARSTON, E. 1977. A comparison of dough preparation procedures for scanning electron microscopy. Food Technol. 31(10):32-36.
YIN, Y. and WALKER, C. E. 1992 Pentosans from gluten-washing wastewater: Isolation, characterizations, and role in baking. Cereal Chem. 69:592-596.

CHAPTER 5
TEMPERATURE -INDUCED CHANGES OF WHEAT PRODUCTS

Peter L. Weegels[1] and Rob J. Hamer[2]
[1]Unilever Research Laboratory, Vlaardingen, the Netherlands and
[2] TNO Nutrition and Food Research Institute, Zeist, the Netherlands.

INTRODUCTION

Temperature plays an important role in the processing of flour. Current processing technology uses a broad range of temperatures: from freezing in frozen doughs, to baking or frying. On a molecular level, conformations of proteins, their polymeric state and their interaction behaviour are affected by temperature. Some of these changes are wanted, others have to be prevented. After an annealing stage starch starts to gelatinise and the water distribution in the dough changes. At higher temperatures, this results in complex changes in which gelatinisation of starch and denaturation of protein are competing. These changes result in large rheological and microstructural changes. The effects of all these changes on product quality will be discussed.

THE EFFECT OF TEMPERATURE ON PRODUCT QUALITY

Temperature, especially during heating, has important beneficial, but in some cases deleterious effects on the quality of wheat products. Temperature induced changes in product quality occur in most stages of wheat processing: drying of wheat, drying of wet gluten, mixing, proofing and baking.

Too moist wheat has to be dried, to prevent sprouting after harvesting and to improve keeping quality. Although severe heating improves the milling characteristics (Mounfield et al, 1944), heat damage will occur under these circumstances. Drying influences the viability of the seeds (Hutchinson and Booth, 1946; Lupano and Anon, 1986; Every, 1987; Schreiber et al, 1981). Prolonged heating of wheat between 60°C and 100°C is reported to be detrimental to loaf volume (Hutchinson and Booth, 1946; Becker and Sallans, 1956; Every, 1987; Hook, 1980; McDermott, 1971; Kent-Jones, 1928; Geddes, 1929, 1930; Mounfield et al, 1944; Schreiber et al, 1981). The most extensive and laborious study ever done on the effects of heat treatment, (1200 loaves of bread were baked) was performed by Geddes (1929, 1930; Fig. 1). Depending on the heating time, he found that temperatures higher than 65°C were detrimental to loaf volume.

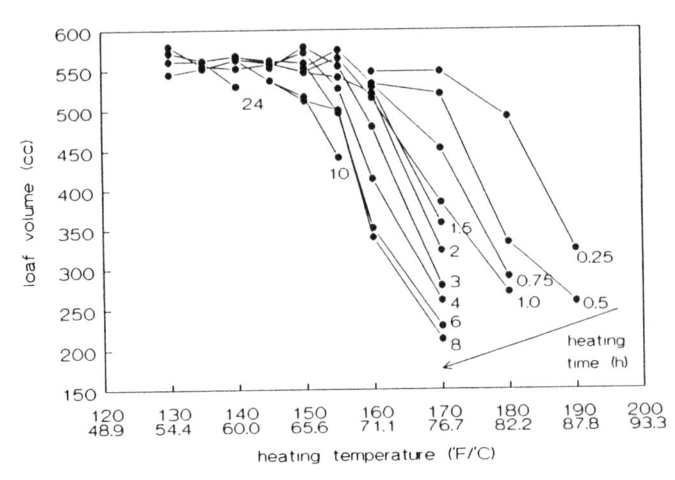

Figure 1. Effect of heating time and temperature of flour on loaf volume (Geddes, 1929; reprinted with permission)

In this early research, Geddes demonstrated that gas retention declined with increased heating of flour, while gas production itself was not affected. He contributed the changes to the inferior gluten quality (Geddes, 1930). Heat damage of wheat occurs more readily at moisture contents above 20% (Becker and Sallans, 1965).

Another part of wheat processing where temperature plays an important role is the production of vital wheat gluten. The applications and use of vital gluten have increased enormously. From 1981 to 1987, gluten production in the former EC has increased from 28,000 tons to 125,000 tons (Gordon et al, 1985). The production in the EU is estimated to be 275,000 tons in 1994. Gluten derives its interesting economic potential from its improvement in dough properties and loaf volume (Gordon et al, 1985; Sarrki, 1979). Thus, strong expensive flours can partly be replaced by weaker ones mixed with gluten. Therefore, gluten has to have a good quality to make its addition financially attractive.

In the production of vital wheat gluten, heat plays also an important role. Gluten is separated from wheat starch by wet processes. In order to prevent deterioration, to enable storage, and to improve handling properties, wet gluten has to be dried. Drying of gluten is considered to be one of the most critical factors determining the quality of vital gluten.

Spray-drying of acid dispersed gluten generally reduce the bread making quality of gluten, except when the inlet temperatures were higher than 135°C and the outlet temperatures lower than 93°C (Lusena and Adams, 1950). Nowadays, wet gluten is industrially dried in ring dryers that are specially designed for drying of gluten. During this drying process inlet temperatures up to 150°C are applied (Barr and Barr, 1975).

A sharp decrease in loaf volume has been found when wet gluten was heated above 60°C (Booth et al, 1980; Pence et al, 1953; LeGrys et al, 1981; Schofield et al, 1983, 1984). An excerpt of their findings is presented in Fig. 2.

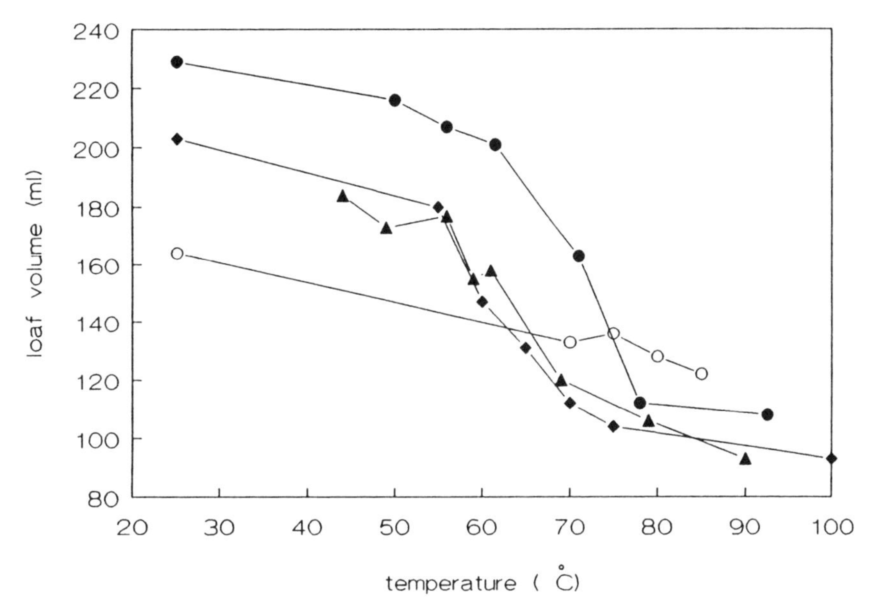

Figure 2. Effect of heating temperature on the bread making performance of flour fortified with freeze dried gluten that had been heated for 5 min at moisture contents between 64 and 67%; (○) Pence et al, 1953; (●) Booth et al, 1980; (♦) Schofield et al, 1983; (▴) LeGrys et al, 1981;reprinted with permission.

Similarly, a decrease in loaf volume was observed when flour was fortified with gluten containing more than 21% moisture that had been heated at 80°C (Weegels et al, 1994a).

The decrease in the baking quality of heated gluten in gluten and starch doughs could be counteracted by adding cysteine (Schofield et al, 1984), sodium metabisulphite or by more intensive mixing (Stenvert et al, 1981). Addition of amylase had no improving effect (Schofield et al, 1984). Since reducing agents and mixing affect the disulphide bonds in gluten, it is likely that disulphide bonds are involved in the loss of baking quality of heated wheat gluten (Schofield et al, 1984; Stenvert et al, 1981).

In each part of the bread-making process temperature plays an important role. During mixing and proofing, moderate heat is used to accelerate yeast growth, to hydrate the proteins, and to improve dough development. During baking of most wheat based products, dough is transformed to the final product. In fact, the differences in baking quality of flour become most clear in this step. In standard baking experiments, dough is shaped to the same size. Prior to baking, bread dough is proofed to the same volume, cake batters are mixed to the same density, biscuits and crackers are sheeted and cut to the same size and puff pastry is sheeted to the same thickness. Despite this uniformity before baking, great differences in volume and size may result after baking when different flours are used. During baking, extremely intricate physical, physico-chemical and biochemical processes occur. This complexity is shown in Fig. 3.

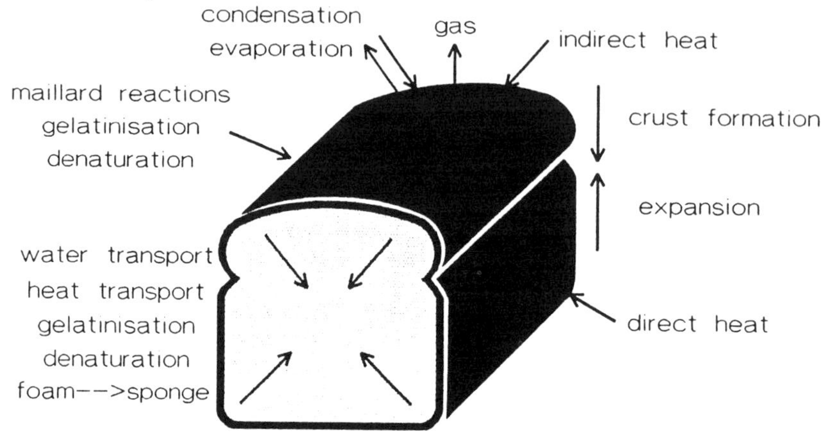

Figure 3. Changes during the transition of a fermented dough to bread during baking.

In a purely physical sense, dough changes from a foam to a sponge structure. The dough expands during heating in the oven (oven-rise), but this expansion is inhibited by crust formation. In a complicated transport of water and gasses (ethanol, CO_2), water vapour condenses initially during baking on the crust, and finally evaporates together with gases. Inside the dough, water is transported from the outer side to the inner side (de Vries et al. 1988). Heat is transported by direct or indirect heat to the loaf. In the crust, starch gelatinises, proteins denature and Maillard reactions occur. Gelatinisation and denaturation also take place inside the loaf, but these changes are very complex owing to the changing moisture content and temperature.

Although this part of the bread making process is of enormous importance to bread quality, only little is known about the chemical and physical changes during bread baking itself. The importance of the heat induced changes of dough, of its components and of their interactions during baking, stresses the need to understand these changes. More insight into and knowledge of the physico-chemical changes that occur during heating, will provide better tools to control, to adapt and to improve the quality of baked products. As these changes are mainly affected by time, temperature and moisture content, some attention will be paid to the reaction kinetics of these changes.

REACTION KINETICS OF HEAT INDUCED CHANGES AND EFFECTS OF MOISTURE CONTENT

In the previous section it was demonstrated that the changes of dough during heating depend to a large extent on three parameters: time, temperature and moisture content. The time-temperature dependence of many reactions, led Arrhenius more than a century ago to the following equation (Equation 1):

$$k = Ae^{\left(-\frac{E_a}{RT}\right)} = d\varphi/dt \qquad (1)$$

Where:

k	= rate constant of the reaction	
A	= constant	
E_a	= activation energy	
R	= gas constant	
T	= temperature	
ϕ	= the ratio of a reaction parameter (*e.g.* loaf volume, extractability, rheological property, etc.) after and before heating	
t	= time	

With this equation, the activation energy of a heat induced change of a parameter may be calculated. The higher the activation energy, the more energy is required to change the parameter by heating and, as a consequence, the later the change will occur during heating. Since the energy of activation is independent of time and temperature, it is more suited to compare the results of various studies. In addition, the energy of activation depends to a large extent on the moisture content of the material studied. In Table I the energies of activation of several parameters are given for some moisture contents.

Schreiber and coworkers (1981) reported values for the energy of activation for loaf volumes from flour from heated wheat of 173 kJ/Mol at 10% moisture and 164 kJ/Mol at 20% moisture. The energy of activation for loaf volume was lower than that for seed viability (208 kJ/Mol to 218 kJ/Mol) and much lower than for the amount of wet gluten that could be isolated from heated wheat (291 kJ/Mol to 298 kJ/Mol). This indicates that loaf volume is more rapidly deteriorated by heating, compared with seed viability and wet gluten content. Up to 20% moisture the energy of activation to reduce loaf volume from heated wheat remained constant at 270 kJ/Mol, but increased at moisture contents above 20% (Becker and Sallans, 1965). At a moisture content of 21.6% the energy of activation was lowered already to 135 kJ/Mol (Becker and Sallans, 1965).

Heating flour at a moisture content of 13.9%, revealed that the rate constant of denaturation, k, increased with temperatures above 63°C. By increasing the moisture content of flour from 4.9% to 13.9% the energy of activation increased. Above 13.9% moisture content the energy of activation decreased rapidly again (data from Geddes, 1929, recalculated by Becker and Sallans, 1956). Comparison of the results obtained with wheat and flour demonstrates that the heat induced changes of wheat and flour differ in their reaction kinetics.

TABLE I

Energy of activation of heating wheat, flour or gluten as determined by several parameters

Parameter	Heated material	Moisture content (%)	Energy of activation (kJ/mol)	Reference
Loaf volume	flour	8.0-19.9	270	Beckers and Sallans, 1956
	wheat	10	173	Schreiber et al, 1981
	wheat	20	164	Schreiber et al, 1981
	flour	21.6	137	Becker and Sallans, 1956
	gluten	64-67	145	Pence et al, 1953
Protein extractability	wheat[1]	17-25	326	Lupano and Anon, 1986
	gluten[2]	64-67	183	Pence et al, 1953
Germination	wheat	10	218	Schreiber et al, 1981
	wheat	20	208	Schreiber et al, 1981
	wheat	17-25	686	Lupano and Anon, 1986
Wet gluten content	wheat	10	298	Schreiber et al, 1981
	wheat	20	291	Schreiber et al, 1981
Glass transition	gluten	12.9	242	Kalichevski et al, 1992a
Tan δ[3] peak temperature	amylo--pectin	indepen-dent	300-740	Kalichevski et al, 1992c

[1] Extractability in 5% NaCl

[2] Extractability in 16% ethanol, 0.083 M acetic acid

[3] Ratio of loss modulus to storage modulus, i.e. viscous to elastic properties

The rate of denaturation is influenced also by the moisture content and the pH at which gluten is heated. At 80°C the rate of denaturation increased when heated at moisture contents above 35% and at a pH higher than 6.5. The energy of activation of wet gluten (64-67% moisture), as determined by loaf volume, was calculated to be 146 kJ/Mol (Pence et al, 1953). This value is comparable with the energy of activation of flour, despite its lower moisture content of 21.6% (Beckers and Sallans, 1956) or flour (Schreiber et al, 1981). Although it is to be expected that the higher the moisture content, the lower the energy of activation, above a moisture content of 20% in flour no large changes are observed in the energy of activation. This can be explained by the observations of various authors that above 20% moisture in starch or gluten, the water activity is higher than 0.9, indicating a fully hydrated system (Fig. 4; Bushuk and Winkler, 1957; Umbach et al, 1992; Weegels et al, 1994a).

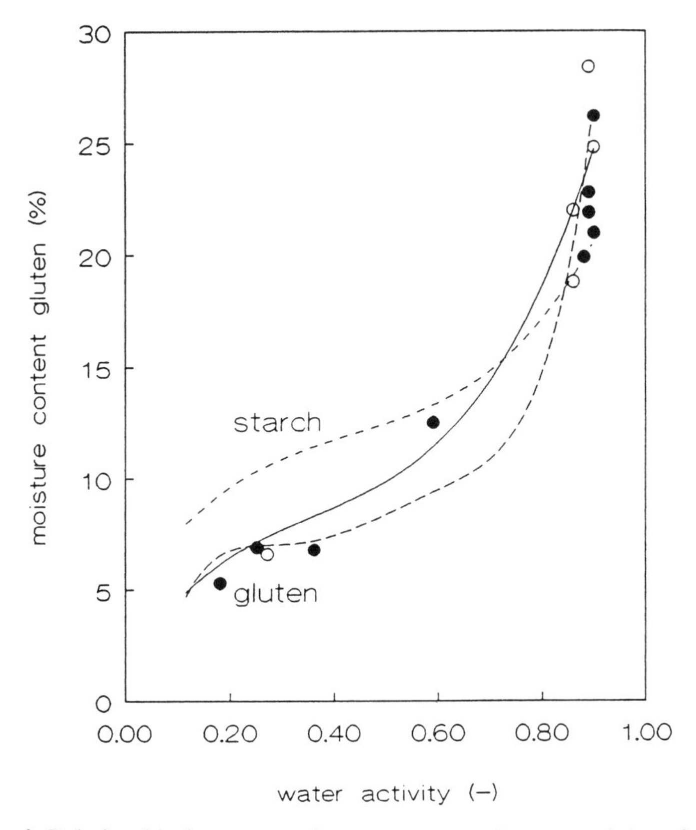

Figure 4. Relationship between moisture content and water activity of gluten and starch; (--) Umbach et al, 1992; (—) Bushuk and Winkler, 1957; (●) unheated gluten, (○) heated gluten, Weegels et al, 1994a; reprinted with permission.

In products with a low moisture content, e.g. biscuits, crackers and extruded flour products, the glass transition is important for the crispness and brittleness of the product (Attenburrow et al, 1992; Slade et al, 1993; Kalichevski et al 1992abc, 1993; Kalichevski and Blanshard, 1992). Glass transitions are sharp increases in the movements between atoms (stretching, bending and rotating). The increase in movements is accompanied by an increase in the amount of energy in a molecule. Since this energy is dependent only on the primary structure, it can be measured with DSC also after denaturation or gelatinisation.

Below the glass transition temperature, the material behaves as a glass. Above the glass transition temperature, these products are deteriorated since they behave as a rubber. The glass transition temperature is strongly affected by plastisizers such as water; the more moisture, the lower the glass transition temperature (Fig. 5).

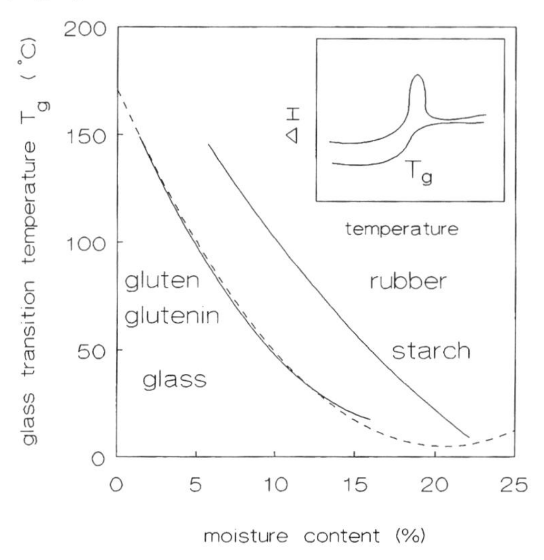

Figure 5. Relationship between moisture content and glass transition of gluten (———; Hoseney et al, 1986), glutenin (---; Cocero and Kokini, 1991) and starch (———; Kalichevski et al, 1993);reprinted with permission.

Above a moisture content of 15%, gluten (Hoseney et al, 1986) or glutenin (Cocero and Kokini, 1991; Kalichevski et al, 1992a; Fujio and Lim, 1989) are in the rubber state at room temperature. Starch behaves as a rubber at room temperature when it contains more than 20% moisture (Kalichevski et al, 1992c, 1993; Kalichevski and Blanshard, 1992). The glass transition of bread closely follows that of gluten (LeMeste et al, 1992). In products with an elevated moisture content, such as bread, the glass transition temperature plays a less important role (Noel et al, 1990, LeMeste et al, 1992). In contrast, the glass transition of the crust is of importance to its crispiness.

In the next two sections, the changes of starch and gluten will be discussed in detail, followed by a section that discusses the changes during heating of starch/gluten mixtures.

During gelatinisation of starch, the starch granules loose their integrity. Differential Scanning Calorimetry (DSC) is generally used to follow the heat induced changes of starch (Liu and Lelievre, 1993; Russell, 1987; Eliasson et al, 1991; Kim and Walker, 1992). The main changes of starch, as determined by X-ray diffraction, DSC and ^{13}C cross polarisation magic anglw spinning NMR (CP-MAS-NMR), occur between 49 and 60°C (Cooke and Gidley, 1992). The cumulative loss of signals obtained by X-ray, DSC and NMR during heating are given in Fig. 6.

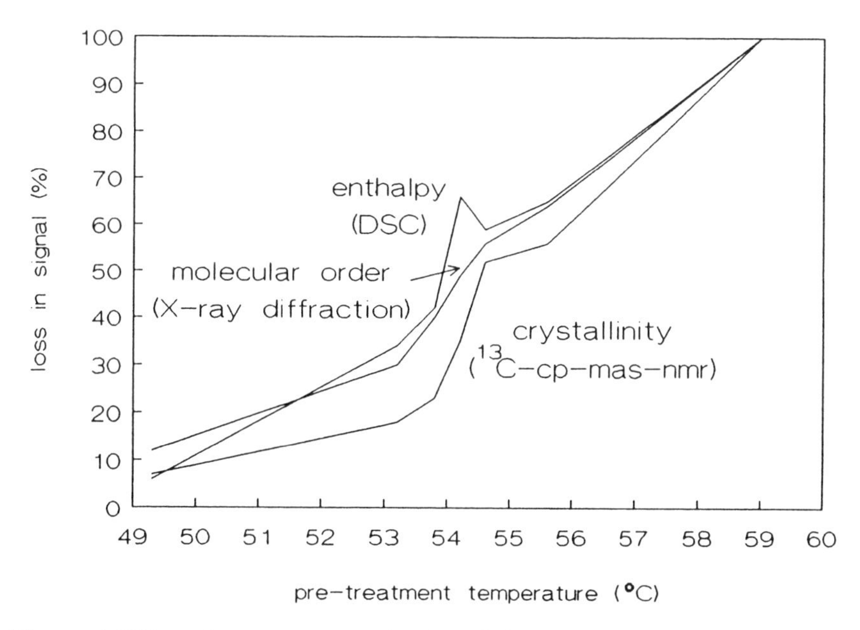

Figure 6. Effect of temperature on the loss of signals measured by DSC, X-ray diffraction and ^{13}C-CP-MAS-NMR from 5% (w/v) starch suspensions (Cooke and Gidley, 1992);reprinted with permission.

The three methods indicate subtle, but completely different, changes during heating. The enthalpy is the energy that is absorbed during heating. X-ray diffraction determines the short range order, i.e. the ordering or interaction of molecules. ^{13}C CP-MAS-NMR is able to determine the overall or longer range order, i.e. the crystallinity. The estimated gelatinisation enthalpy values for starches with an hypothetical 100% molecular order (25-41 J/g) resembled the enthalpy values measured for debranched glycogen (34-35 J/g; more than 85% crystalline and more than 90% molecular order) more closely than the hypothetical enthalpies for 100% crystalline order (45-70 J/g). From this comparison and from Figure 6, it can be inferred that the change of enthalpy closely follows the loss of molecular order and less so that of crystallinity (Cooke and Gidley, 1992). This indicates that changes of enthalpy of starch are governed more by changes in the short range molecular order and less by changes in the long range crystalline order.

In Fig. 6 subtle differences are shown, but very large differences can be observed when changes determined by other techniques are compared with the changes of enthalpy (Tester and Morrisson, 1990; Cooke and Gidley, 1992), such as the storage modulus, the phase angle (i.e. the ratio of viscous over elastic properties (Eliasson, 1986), solubility of amylose and swelling of the starch granules (Ellis et al., 1989) and birefringence (Liu and Lelievre, 1993; Fig. 7).

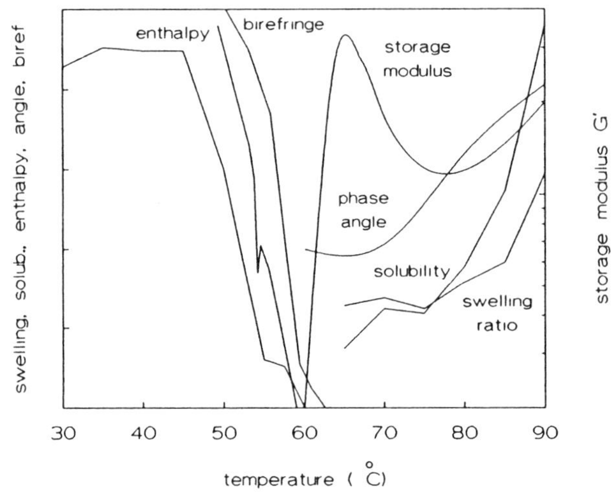

Figure 7. Comparison of the changes of enthalpy (Cooke and Gidley, 1992; Tester and Morrisson, 1990), birefringence (Liu and Lelievre, 1993), storage modulus and phase angle (Eliasson, 1987), and starch solubility and swelling ratio (Ellis et al, 1989) upon heating of 5-10% (w/v) starch suspensions ;reprinted with permission.

Care has to be taken to compare results obtained in different studies in one figure since different conditions were used. Nevertheless, Fig. 7 demonstrates that functional changes, such as rheological changes, do not occur at the same temperatures of the molecular or physico-chemical changes. Therefore, the methods will not be equally effective in explaining the deterioration of product quality during heating.

Another interesting change may be determined by ^{17}O CP-MAS NMR. In this technique the τ_2 relaxation time is a measure for the interaction between water and starch. The shorter the relaxation time, the stronger the interaction (Fig. 8).

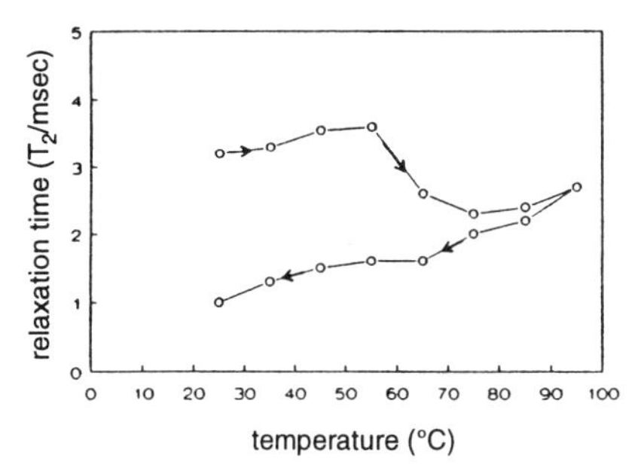

Figure 8. Effect of heating and cooling on the content of free water as determined by the relaxation time T_2 of ^{17}O-NMR measurements. The larger T_2 the more free water (Chinachoti et al, 1991);reprinted with permission.

It can be seen from Fig. 8 that between 55 and 65°C the relaxation time decreases and more water becomes bound. This effect is not reversed upon cooling. The increased interaction of starch with water during and after heating will certainly affect the other water dependent changes during heating.

The sequence of the transitions may give a better insight into the fundamental changes that occur. In summary, methods to determine changes of starch characteristics during heating differ not only in their sensitivity but also in the type of changes they detect (Fig. 9).

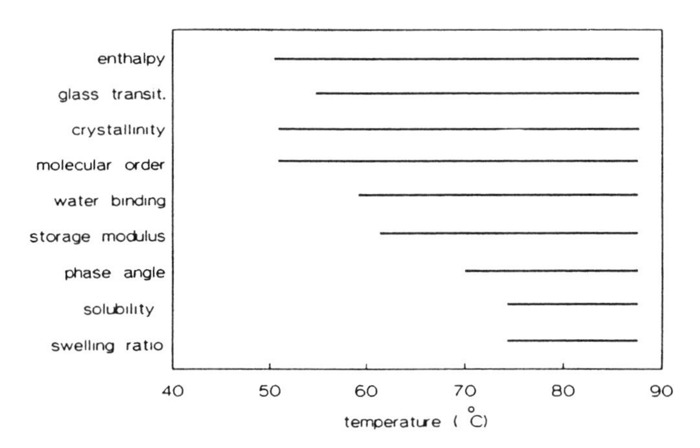

Figure 9. Summary of the sequence in changes determined in the studies mentioned in Fig. 7 of starch suspensions during heating.

During the first stage in the gelatinisation, molecular order and crystallinity are lost, accompanied by an increase in enthalpy. Directly after these changes, birefringe of the starch kernels is lost. The changes result into a more strong binding of water. Next, the storage modulus of starch suspensions increases, followed by an increase in the phase angle, i.e. the proportion of viscous properties. Finally, the solubility of starch increases and the starch kernels swell. Thus, water binding, solubility and swelling contribute to the rheological properties in different and, as yet, unexplained ways.

PHYSICO-CHEMICAL CHANGES OF GLUTEN PROTEINS DURING HEATING

Temperature Induced Changes of Rheological and Thermal Properties

In contrast to starch, the physico-chemical changes of gluten proteins during heating are less well investigated. As discussed at the beginning of this chapter, when wet gluten is heated above 70°C or when the moisture content during heating is above 20%, bread making quality is deteriorated.

Elasticity and viscosity determine the processing properties of a dough. Since gluten is one of the main components contributing to these properties, rheological methods are often used to determine the changes of the gluten quality. Compared to native wet gluten, no changes were detected in the Farinograph mixing curves of gluten-water mixtures when freeze-dried gluten had been heated at 70°C or less than 2.5 h at 80°C. At longer heating times or at higher heating temperatures gluten lost its cohesiveness and no meaningful Farinograms could be made (Doguchi and Hlynka, 1967). Gluten-stretching tests were more sensitive in detecting heat damage to dry gluten since heating for 1 h at 70°C showed a deviation in extension of gluten (Doguchi and Hlynka, 1967).

Using small deformation dynamic rheological measurements, it can be shown that the modules of gluten changes continuously upon heating (LeGrys et al., 1981; Fig. 10).

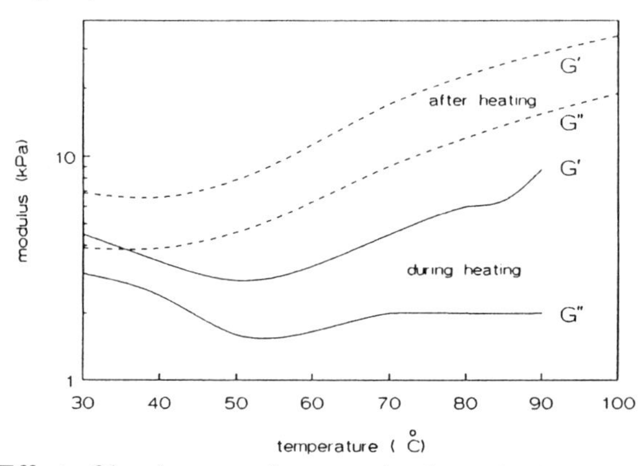

Figure 10. Effect of heating wet gluten on the dynamic rheological properties during and after heating (LeGrys et al, 1991 ;reprinted with permission).

When these properties were measured during heating, first the storage and loss moduli decreased, probably because hydrogen bonds were weakened. Above 55°C, the moduli increased, especially the loss modulus, thus increasing the proportion of viscous properties of gluten. This increase is partly due to the gelatinisation of residual starch. When the dynamic rheological properties of heated gluten are determined after cooling, both the storage and loss modulus increased above 55°C (LeGrys et al., 1981). The storage modulus was inversely related to loaf volume (LeGrys et al, 1981). In contrast, Dreese and coworkers (1988ab) reported that after heating a gluten-water dough to 90°C and cooling to 30°C, no large changes of the dynamic rheological properties were detected. The small change that occurred, was demonstrated to be caused by the residual starch in the gluten. The authors proposed that differences in dynamic rheological properties owing to heating were eliminated since the dough with denatured gluten was mixed to complete development (Dreese et al, 1988ab). The discrepancy between the results from LeGrys and colleagues (1981) and Dreese and coworkers (1988ab) may be explained by differences in the way gluten were mixed.

By stretching gluten that had been heated at temperatures above 55°C, the viscosity and elasticity of gluten increased sharply. Viscosity increased relatively more than elasticity. From 30°C to 50°C the viscosity decreased, probably owing to the reduction of the number of cross-links (Bale and Muller, 1970). On the contrary, elasticity was not affected in this temperature range (Bale and Muller, 1970; Hermansson, 1983). By using large deformations in creep measurements, the apparent viscosity of dough containing 4% (flour basis) gluten that had been heated for 30 min at 80°C at moisture contents above 6.7%, increased gradually (Weegels et al, 1994a). In creep measurements in the Glutograph, heat damaged gluten needed a longer sheartime to reach 800 BU (Weipert and Gerstenkorn, 1988), indicating a higher degree of viscosity.

The Brabender Extensograph Maximum Resistance and Extensibility decreased when dough fortified with gluten that had been heat treated at moisture contents exceeding 20%. The Extensograph Resistance after 50 mm (R_1) paper transport increased gradually with increasing moisture content of the gluten during heat treatment (Weegels et al, 1994a). This is consistent with the finding that the R_1 values of dough, containing a range of commercial wheat gluten samples, were negatively related to the loaf volume (Weegels and Hamer, 1989).

Above a heating temperature of 60°C, the elasticity, as measured by the relaxation time, increased sharply (Hermansson, 1983a). Hermansson concluded from her rheological, scanning electronmicroscopy and DSC findings that the major changes of heated wet gluten are due to the gelatinisation of starch (Hermansson, 1983a).These results indicate that gluten is a rubber that may become cross-linked similar to the vulcanisation of normal rubber.

In general, proteins show marked thermal transitions during denaturation (Privalov and Khenchinashvili, 1974; Foegeding, 1988). Apart from the glass transition, gluten proteins display no clear thermal transition upon heating. The enthalpies of the thermal transitions of isolated gluten are about 100 times lower compared with other proteins (Eliasson and Hegg, 1980; Lupano and Anon, 1986; Hoseney et al, 1986; Hermansson, 1983a; Arntfield and Murray, 1981; Fujio and Lim, 1989; Schofield et al., 1984). This may be the result of the reformation of hydrophobic and hydrophilic bonds immediately after they are disrupted. Thus, only a small net thermic flow will be measured by DSC (Arntfield and Murray, 1981; Foegeding, 1988). The major difference between gluten and other proteins is its high glutamine content. This amino acid is responsible for substantial hydrogen bonding. One may speculate that there is a hydrogen bond interchange during heating that is responsible for the deterioration in quality. Furthermore, gluten consists of a mixture of 100 to 150 different proteins. When these proteins denature at different temperatures, there will be no clear thermal transition.

Temperature-induced Changes of Extractability

One of the most marked features of protein denaturation is its decreased solubility and protein extractability is often taken as a measure for the degree of denaturation. Since gluten is unextractable in water, but partly extractable in urea, sodium dodecyl sulphate, guanidine HCl or acetic acid, most extractability studies have been carried out with these solutions. With increasing heating time, moisture content and temperature, gluten becomes less soluble in urea, SDS, guanidine HCL or acetic acid (Pence et al, 1953; Lupano and Anon, 1986; Booth et al, 1980; Schofield et al, 1983; Schofield et al, 1984; Jeanjean et al, 1980; Weegels et al, 1994a), water (Hutchinson and Booth, 1946), or $MgSO_4$ or KI (Geddes, 1930). The decrease in extractability was highly correlated with the decrease in loaf volume (Geddes, 1930; Pence et al, 1953). The baking test was more sensitive to detect denaturation since changes were observed earlier than with the solubility determination (Pence et al, 1953; Geddes, 1930; Hutchinson and Booth, 1946).

Several studies point out the differential denaturation behaviour of wheat proteins. One of the earliest extractability studies reported a substantial decrease in extractability of the globulin fraction (i.e. the proteins extractable in $MgSO_4$ or water) from wheat or flour that had been heated for more than 3 h at 70°C (Geddes, 1930; Berliner and Rüter, 1928). As determined by SDS-PAGE, the extractability of the albumin and globulin fraction was the first to be reduced when a flour slurry (Wrigley et al, 1980) or gluten were heated (Jeanjean et al, 1980). A test to determine the degree of denaturation was developed that makes use of the decreased solubility of the saline soluble proteins (McDermott, 1971). When gluten was heated for 30 min at 80°C, no large changes of extractability of albumin/globulin could be detected (Weegels et al 1994a).

Extractability of glutenin is reduced by 32% when flour is heated for 10 hours at 80°C (Herd, 1931) and by up to 40% when it is heated for 15 hours at 100°C (Berliner and Rüter, 1928). Booth (1980), Schofield (1983), Weegels (1994ab) and co-workers found that the decreased extractability of gluten is mainly accounted for by the decreased extractability of the glutenin. An example is given in Fig. 11.

Changes of the extractability in SDS (in which glutenin is partly extractable) occurred already when wet gluten was heated for 5 min at 60°C, whereas the extractability in propan-1-ol (in which glutenin is not extractable) was detected only after heating at 75°C for 5 min (Schofield et al, 1983). After applying a temperature of 90°C, glutenin became almost completely unextractable in 4.5 M urea (Booth et al, 1980) or SDS (Schofield et al, 1983).

In contrast, when a suspension of wheat gluten in SDS was heated at 100°C for 15 min, an increase in extractability was observed. Addition of 2-mercaptoethanol increased protein extractability equally in heated and non-heated gluten suspensions in SDS (Rizvi et al, 1980). From these data it can be inferred that SDS prevents the coagulation of the proteins during heating.

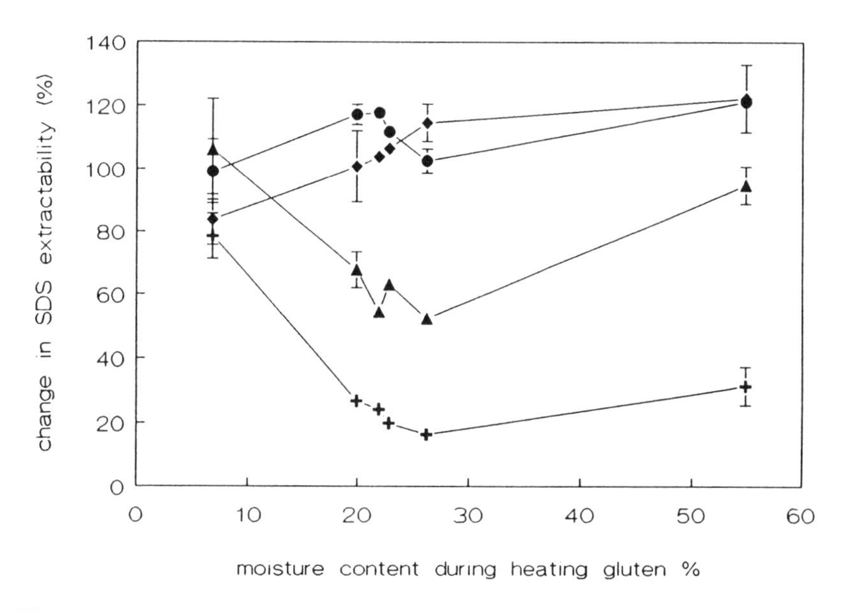

Figure 11. Extractability of high M_r glutenin aggregates (+; $M_r > 2\times10^6$), low M_r glutenin aggregates (▲), gliadin (♦) and albumin/globulin (●) from gluten that had been heated for 30 min at 80°C at different moisture contents, and that had been extracted subsequently by 1.5% (w/v) SDS and separated by gel filtration chromatography on Sepharose CL-6B (Weegels et al 1994a).

The protein extractable in aqueous ethanol solutions is often referred to as the gliadin fraction. In addition to these monomeric proteins, it was recently demonstrated that the fraction contains also some polymeric glutenin proteins (Huebner and Bietz, 1993). The extractability in 60-65% ethanol solutions of gluten or gliadin that had been heated for at least 30 min at 70°C was reduced only by 7-15% (Alsberg and Griffing, 1927; Cook, 1931). When flour was heated below 70°C, Geddes found an increase in extractability of proteins in 70% alcohol. Heating at 77°C showed virtually no change of extractability (Geddes, 1930). Similar results were obtained in later studies (Pence et al, 1953; Schofield et al, 1983; Weegels et al, 1994a). In contrast, the extractability in aqueous ethanol of proteins from flour that had been heated under more severe heating conditions (8 to 10 h between 70 and 96°C) was reduced by 30 to 44% (Berliner and Rüter, 1928; Herd, 1931). In baked bread, increased baking time resulted in a reduced protein extractability in 80% ethanol (Westerlund et al, 1989).

The rate of denaturation, as determined by the loss of extractability of gliadin, was three times less than the rate of denaturation of gluten (Pence et al, 1953). From these findings it can be inferred that the rate of denaturation is higher for glutenin than for gliadin.

The changes in extractability differed between the groups of gliadin proteins. Upon heating of wet gluten at 100°C (Booth et al, 1980; Schofield et al, 1983), after baking (Pomeranz, 1988), after drying pasta products at 90°C (Feillet et al, 1989), or after heating a flour slurry at 130°C (Wrigley et al, 1980), α-, ß- and γ-gliadins, but not the ω-gliadins, became less extractable, as was shown by electrophoresis (Booth et al, 1980; Pomeranz, 1988; Feillet et al, 1989; Wrigley et al, 1980) and RP-HPLC (Pomeranz, 1988). This differential behaviour may be explained by the absence or reduced number of sulphydryl groups in the ω-gliadins. In the other gliadins the sulphydryl groups were able to form intermolecular disulphide bridges, making the proteins unextractable (Booth et al, 1980; Schofield et al, 1983).

From the extractability studies it becomes clear that glutenin, albumin and globulin are most sensitive to heat treatment. Whether the decrease in extractability is owing to an increased complexation, aggregation, cross-linking or denaturation of the proteins per sé, is not clear. Another complicating factor is the entrapment of one type of proteins (e.g. gliadin) in an aggregate or complex of insoluble proteins from another type (e.g. glutenin). As Alsberg and Griffing state: "Whether the gliadin actually suffers a change of solubility or whether it merely becomes more difficult to extract, it is difficult to say." (Alsberg and Griffing, 1927; p420). Unfortunately 60 years later this ambiguity still is not clarified.

Changes in Size Distribution

Apart from denaturation, the decrease in extractability may be caused by an increased association of the proteins. Changes of the degree of association in extracted protein can easily be observed by gel permeation chromatography. The amount of high M_r glutenin complexes extractable in urea, SDS or acetic acid/urea/cetyl trimethyl ammonium bromide buffers with M_r's larger than 7×10^7 (Booth et al, 1980), 1×10^5 (Schofield, 1983,1984), 8×10^5 (Feillet et al, 1989), 2×10^6 (Weegels et al, 1994a) or 1×10^8 (Weegels 1994b) decreased more than lower M_r glutenin aggregates (Booth et al, 1980; Schofield et al, 1983; Weegels et al, 1994ab) or gliadin (Feillet et al, 1989).

Changes in the degree of association in the unextracted proteins are not easily measured directly, but can be calculated from rheological measurements by using the Flory equation, as was done by Bale and Muller (Muller, 1969; Bale and Muller, 1970; Equation 2).

$$f=RT\rho\frac{v_2^{1/3}}{M_c}(\alpha-\frac{1}{\alpha^2}) \qquad (2)$$

Where:

f	=	tensile stress
R	=	gas constant
T	=	temperature
ρ	=	density of the dry material
ν_2	=	volume fraction
α	=	stretched length of the swollen polymer divided by the initial length
M_c	=	average M_r of the chain between cross-links

The average number of cross-links per unit mass is given by $1/(2M_c)$. After measuring the tensile stress, the number of cross-links may be calculated. The average number of cross-links increased three to four times when gluten had been heated between 70°C and 80°C (Bale and Muller, 1970; Muller, 1969). No other attempts were found to determine changes of M_r distribution after heating by this approach. Dynamic rheological measurements seem to be promising since they can monitor changes of the size of the SDS unextracted glutenin aggregates during dough processing quite well (Weegels, 1994).

With Circular Dichroism (CD) measurements it is possible to determine the content of secondary structures on a molecular level such as α-helix, ß-sheet and ß-turn, and random elements in a protein. Unfortunately, only proteins in solution can be measured. Upon heating, proteins may become unextractable and the change of conformation cannot be measured. This is especially a problem with gluten that is unextractable in water. Upon heating α-gliadin (Kasarda et al, 1968; Tatham et al, 1987) or low M_r glutenin (Tatham et al, 1987) up to 80°C, α-helix and ß-sheet contents decreased both from 12% to 8% (Kasarda et al, 1968; Tatham et al, 1987). After cooling to ambient temperatures, these decreases appeared to be reversible (Kasarda et al, 1968).

Glutenin can only partly be brought into solution by salts that are known to denature proteins (Field et al, 1987; Cluskey and Wu, 1971). In contrast, the use of organic solvents (Tatham et al, 1987), SDS (Stewart II et al, 1974; Takeda et al, 1988) or urea (Takeda et al, 1988) only to a minor extent interfere with the conformation of several proteins. This opens possibilities to determine the conformation of heated gluten extracted in these solutions. Large changes of secondary structure were observed in glutenin that had been extracted in SDS from gluten that had been heated for 30 min at 80°C and that had been separated from the other protein fractions by gel permeation chromatography (Weegels et al, 1994b). When gluten was heated at moisture contents higher than 20% the proportion of α-helical structure decreased and the proportion of random elements increased (Fig. 12).

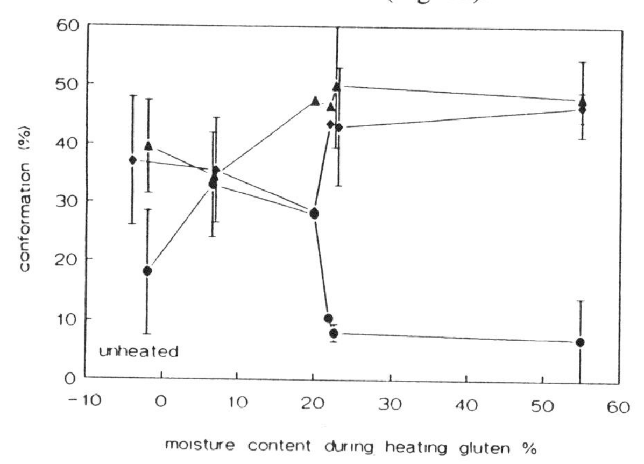

Figure 12. Change in secondary structure as predicted by circular dichroism of high and low M_r glutenin from gluten that had been heated for 30 min at 80°C at different moisture contents, and that had been extracted subsequently by 1.5% (w/v) SDS and separated by gel filtration chromatography on Sepharose CL-6B. (●) α-helix; (▲) ß-sheet; (■) ß-turn and random coil (Weegels et al 1994b).

No such changes were observed in the gliadin or albumin/globulin fraction (Weegels et al, 1994b).

With ^{13}C and ^{13}N NMR spectra, intensity changes of peaks and chemical shifts were detected when gliadin was heated (Baianu et al, 1982). Baianu and coworkers concluded from their data that, upon heating, inter-molecular hydrogen bonds were broken and intra-molecular hydrogen bonds were formed. Some of the aromatic amino acids were restricted in their motion after heating, but others had a greater mobility (Baianu et al, 1982).

In the near-UV region an increase in intensity of absorption of aromatic side chains in the protein was detected at elevated temperatures (Tatham et al, 1987; Kasarda et al, 1968). In denaturing solutions a decrease in intensity was found, however (Kasarda et al, 1968; Field et al, 1987). These changes are difficult to interpret, but Kasarda et al (1968) speculated that they were caused by aggregation since they observed an increase in intensity of absorption of the aromatic side-chains upon salt induced aggregation of gliadin. This indicates that a more rigid environment of the aromatic side-chains causes a more intense absorption, whereas in denaturing solvents aromatic side-chains of the protein was in a less rigid hydrophobic environment. As a consequence, their intensity would be decreased.

Another method to analyze changes in conformation owing to heating is CP MAS NMR. The advantage of this method is that, unlike other methods, proteins do not need to be in solution. No changes were detected, however, when glutenin or gliadin are heated (Ablett et al, 1988).

Gluten consists of more than 25% glutamine residues. This amino acid has the capability to interact with other glutamine amino acids by double hydrogen bonds. Individually, hydrophobic and hydrophilic interactions are small in force, but the regular and repetitive character of these bonds (Sugiyama et al, 1985; Shewry et al, 1992) enable a strong interaction unique for gluten proteins (Schofield and Booth, 1983; Wall and Beckwith, 1969; Belitz et al, 1986). Strong hydrophobic or hydrophilic interactions may increase the size of protein aggregates and the formation of a gel can become evident (e.g. Booth et al, 1983; Schofield et al, 1983, 1984).

In proteins the surface hydrophobicity generally increases upon heating (Tanford, 1980; Wicker et al, 1986; Voutsinas et al, 1983; Matsudomi et al, 1982; Deshpande and Damodaran, 1989). A reduction in surface hydrophobicity may be observed, when the increased hydrophobicity leads to aggregation, however (Mangino, 1988; O'Neill and Kinsella, 1988). Therefore, the measurement of hydrophobicity may give valuable information about the mechanism of heat induced changes.

According to Tanford the increase in hydrophobicity owing to heating is caused by the exposition of hydrophobic groups to the outside of the molecule (Tanford, 1980). This increases the possibilities of ligands to bind in the hydrophobic region. Changes of hydrophobicity can be monitored by using hydrophobic ligands such as anilino naphtalene sulphonic acid (ANS) that has a strong fluorescence in a hydrophobic environment. An increase in fluorescence of ligands was observed even before an increase in rigidity of a fish protein gel is observed (Wicker et al, 1986). When gluten had been heated a strong decrease was observed in the number and the binding constant of ANS molecules that were bound. The decrease in hydrophobicity was probably caused by a strong aggregation (Weegels et al 1994b; Fig. 13).

Indirect evidence of the importance of hydrophobic properties during heating is ample. Succynilation and deamidation prevent the formation of a gel when gluten is heated (Anno, 1981). Since these modifications decrease the hydrophobic character of gluten and its ability of extensive hydrogen bonding, these properties are important for gel-formation. Salts that are able to break hydrophobic and hydrophilic bonds, such as SDS, urea and guanidine Hcl, reduce the hardness of a gel from heated gluten, when present during heating (Anno, 1981). Feillet and coworkers postulated that the boiling quality of pasta products is determined by its SH/SS content and by the formation of hydrophobic bonds after boiling. Indirect evidence was derived from physico-chemical studies, such as protein solubility, SDS- and lactate- PAGE (Feillet et al, 1989; Jeanjean et al, 1980) and size exclusion HPLC (Feillet et al, 1989).

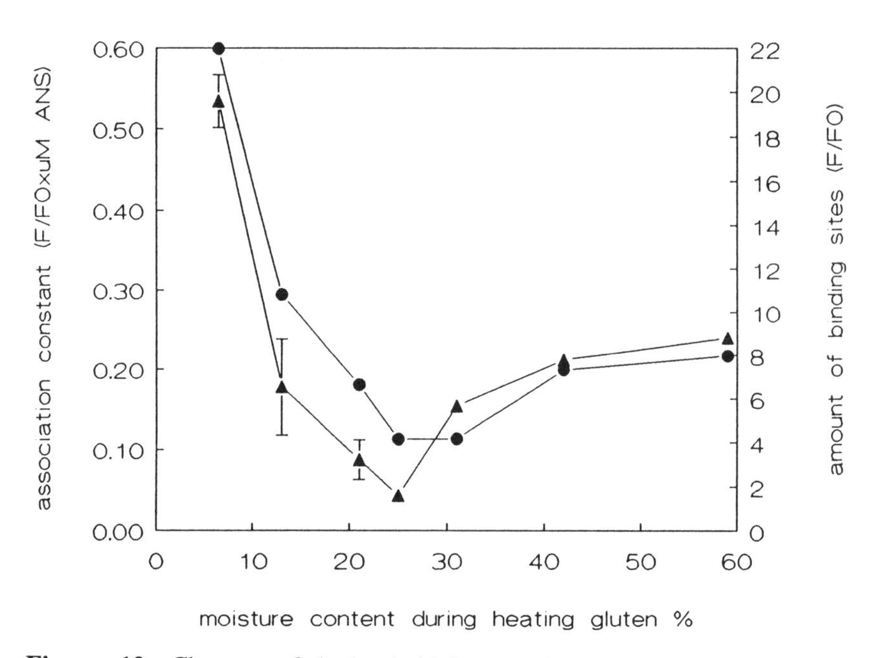

Figure 13. Changes of hydrophobicity as determined by the relative fluorescence (F/F0) of 8-anilino-naphthalene sulphonic acid bound to gluten with different moisture contents that had been heated for 30 min at 80°C. (▲) apparent association constant; (●) apparent number of binding sites (Weegels et al, 1994b).

Another method to determine changes of hydrophobicity during heating is electron spin resonance (ESR). The electron spin resonance of spin probes is influenced by the micro-environment of the probe. Thus, the partitioning of a spin probe in a hydrophilic and hydrophobic regions can be determined. The amount of the probe 2,2,6,6-tetramethylpiperidinyloxy (TEMPO) that was bound in a hydrophobic environment, increased after heating up to 95°C (Pearce et al, 1988).

Changes in the interaction of protein and water give also important information on the structure of protein during heating. The mobility of protons in protein or protons of water in the environment of parts of a protein may be determined by pulsed NMR. A decreased mobility of protons in proteins may be an indication for an increased aggregation, whereas a decreased mobility of water protons indicates a more strong binding of water to protein. With pulsed NMR the ^1H spin-spin relaxation could be divided into three components with a fast (2 ms), medium (28 ms) and a slow (69 ms) relaxation. Fast relaxation times were contributed to protein protons. Wet gluten that had been heated at 50°C or 70°C and cooled to room temperature showed no change of the fast relaxation of the protein protons (LeGrys et al, 1981; Ablett et al, 1988). The relaxation time of the medium relaxation component increased after heating. There appeared to be a water transfer to the medium component. The water had become a part of the bulk water (LeGrys et al, 1981). The mobility of peptide chains increased with increasing temperature, the sharpest increase being above 60°C. When cooled to ambient temperature no changes could be detected. It was concluded that labile non-covalent interactions were replaced by covalent disulphide bonds, but with this method no permanent changes of conformation were detected also (Ablett et al, 1988).

General conformational changes of protein can be measured with Infra Red spectra (Rudzik et al, 1987), but no literature was encountered using this technique for gluten.

Chemical Changes

In 1930 it was suggested already by Geddes that oxidation occurred when flour was heated. A too severe oxidation of some unknown constituents was assumed to be deleterious for the baking quality (Geddes, 1930).

The increase in amount of SDS-unextracted high M_r glutenin protein aggregates may be caused by an increase in covalent cross-links. The most important covalent cross-links in gluten are the disulphide bridges. For kinetic reasons heating will accelerate the formation of these S-S bridges. Once covalent coupling by S-S bonds has occurred, physical aggregation is difficult to reverse.

Heating a flour dough in an oven at 130°C decreased the amount of free sulphydryl groups from 16.8 μmol/g protein to 14 μmol/g protein and of reactive sulphydryl groups from 11.7 μmol/g protein to 9.8 μmol/g protein.

Concomitantly, mixing time increased and the rate of dough breakdown decreased. Addition of reduced glutathione counteracted these changes (Okada et al, 1987). When gluten was heated with a moisture content exceeding 20%, the total sulphydryl content decreased (Weegels et al, 1994b). Both studies indicate the importance of disulphide bonds during heating for the bread making quality of gluten. Schofield and coworkers (1983, 1984) did not find any change of the total amount of free sulphydryl groups during heating wet gluten. They found that the amount of free SH groups in the SDS extracted protein decreased, whereas the amount of free sulphydryl groups in the SDS unextracted protein increased (Fig. 14). They concluded that, upon heating, a disulphide/sulphydryl exchange, instead of a formation, had taken place (Schofield et al, 1984). When their results are recalculated to relative proportions on protein basis, the concentration of sulphydryl groups in SDS extracted and unextracted protein both decrease. This indicates that, upon heating, more disulphide bridges are being formed.

Glutenin subunits are known to be covalently linked by disulphide bridges to form large glutenin polymers. Therefore, it is acceptable that glutenin proteins are involved in the disulphide formation. By labelling sulphydryl groups with radio-active iodoacetamide and separating the SDS-soluble proteins, Schofield and coworkers found that the sulphydryl groups of glutenin were more affected by heating than those of gliadin. The radio active labelled glutenin became incorporated in the very high M_r (probably larger than 5×10^6) SDS-extracted proteins or became unextractable in SDS. They postulated that, upon heating, the proteins unfolded, thus making disulphide/sulphydryl interchanges easier. Owing to this interchange, proteins were fixed in the denatured state (Schofield et al, 1983). Similar results were obtained by labelling gluten with the fluorescent sulphydryl reagent monobromobimane: the labelling was reduced in glutenin from heat treated gluten, but not in gliadin (Weegels, 1991). In SDS extracted proteins from heated and unheated gluten, the sulphydryl content remained equal. The ratio of sulphydryl group to total half cystine in SDS unextracted protein decreased dramatically from 11×10^{-3} to 0.9×10^{-3} upon heating, however (Weegels et al, 1994b). The differential extractability behaviour of various groups of gliadins is related also to their sulphydryl content, as discussed earlier (Booth et al, 1980; Schofield et al, 1983; Pomeranz, 1989; Feillet et al, 1989).

Jeanjean and coworkers found a decrease in extractability of the proteins with a low M_r, such as globulin and albumin. Since these proteins became extractable only after addition of 2-mercaptoethanol, it was concluded that they participate in the formation of an unextractable protein complex by disulphide bonds (Jeanjean et al, 1980).

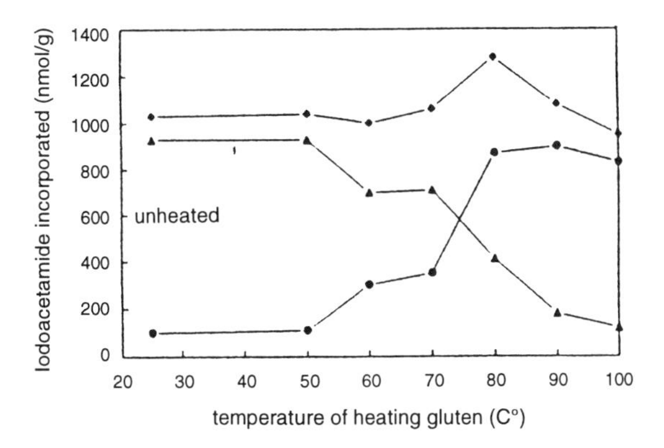

Figure 14. Effect of temperature on the sulphydryl content as determined by ^{14}C-iodoacetamide labelling of SDS-extracted gluten (▲), SDS-unextracted gluten (●) and of total gluten (◆) that had been heated for 5 min while fully hydrated (Schofield et al, 1983);reprinted with permission.

Mechanisms of Denaturation of Gluten Proteins

Information on heat induced changes of gluten is too scarce to obtain a clear insight into the mechanisms that ultimately result in functional changes. When the sequence of these changes would be known, this would provide a better insight into the cause and effect relationships that might explain the effect on product quality.

For other proteins, hypothetical models were developed describing the sequence of heat induced changes of proteins. Catsimpoolas and Meyer (1970) when studying soya bean globulin denaturation, postulated that first the quaternary structure is disrupted. At this stage hydrophobic interactions are important since they become more strong upon heating. Next, covalent bonds are formed. The third step in this sequence is the restorage of hydrogen bonds after cooling. Experimental evidence demonstrated that the state before the covalent bond formation, was still reversible. Excessive heating caused irreversible changes (Catsimpoolas and Meyer, 1970). The observation that the denaturation of soya bean proteins depends on the hydrophobicity of the alcohol in which they are denatured, support the hypothesis (Fukushima, 1968).

The sulphydryl groups in glutenin are located at the C and N terminal end of the high M_r glutenin subunits (Shewry et al, 1992). The location and the relatively small number of sulphydryl groups present make it unlikely that disulphide formation is the first step or the driving force in the rather random process of denaturation. That disulphide cross-linking is not the cause of the physico-chemical changes during heating of whey proteins, was clearly shown by Li-Chan (1983), although the extractability of the heated whey protein concentrate was statistically correlated with its sulphydryl content. When 2-mercaptoethanol was used to extract the proteins, the extractability increased, but the relationship between the sulphydryl content and the extractability remained the same. There appeared to be no causal relationship between sulphydryl content and extractability. It was inferred that the changes of sulphydryl content are only a consequence of the changes, but not the cause (Li-Chan, 1983). Similar relationships have been postulated for denaturation of other proteins (see review of Kilara and Sharkasi, 1986) such as: milk proteins (Kalab et al, 1973; Kalab and Emmons, 1972; Beveridge et al, 1984), egg-white (Mangino et al, 1988), and soya bean and albumin protein (Beveridge et al, 1984). This does not mean that disulphide bonds are not important in the denaturation. On the contrary: since covalent cross-links are not easily reversible they are able to fix the denatured state of the proteins. A summary of the tentative sequence of events during heating gluten is given in Fig. 15.

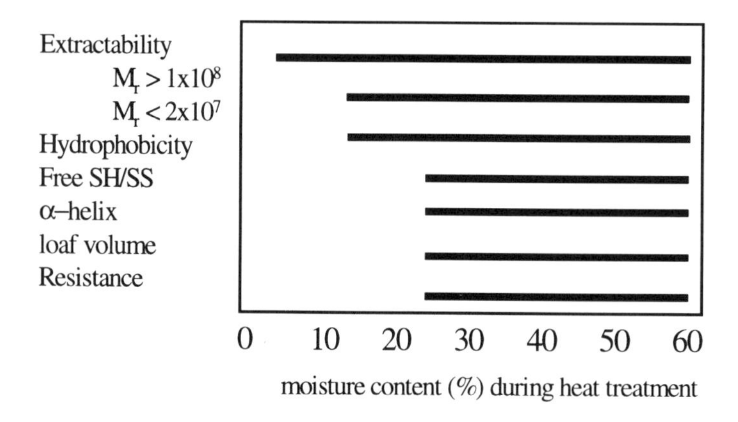

Figure 15. Summary of the sequence in changes of gluten that had been heated at different moisture contents for 30 min at 80°C (Weegels et al, 1994ab).

Decreases of extractability of the largest molecular mass glutenin aggregates and in hydrophobicity are observed already when gluten is heated at moisture contents below 20%. Conformational changes, changes of sulphydryl groups and changes of functional properties such as loaf volume or rheological properties are observed only when gluten is heated with a moisture content more than 20% (Weegels et al, 1994ab). As with starch, not all changes of properties of gluten after heating are directly linked to the functional changes.

INTERACTIONS: HEAT INDUCED CHANGES OF GLUTEN/STARCH MIXTURES

As discussed in the introduction, the changes of the dough during baking are very complex. First of all, there are temperature and moisture gradients that affect strongly the reaction kinetics. Secondly, dough is complex a mixture of mainly starch, gluten and water. As demonstrated earlier, the individual components undergo specific changes during heating. In the dough these changes are mutually dependent. Therefore, it is not surprising that virtually no physico-chemical studies have been performed that describe the changes during baking and much is matter for speculation.

Starch and gluten probably compete for water during heating (Ghiasi et al, 1982), therefore, it is of interest to determine the changes of water binding of the individual components during heating. The water distribution in dough is given in Table II.

TABLE II

Water distribution in dough before and after heating

Fraction[1]	Dough before heating[1]		Dough after heating[1]
Non-starch polysaccharides	23[2]	15[3]	23
Protein	31	28	
Granular starch	26	33	
Damaged starch	19	23	77

[1] Data represent percentage of water bound.[2] Bushuk, 1966; [3] Calculated from Bloksma and Bushuk, 1988

In the dough, most of the water is bound to protein: 28 to 31%. After heating, most of the water was probably bound to the damaged starch, as was speculated by Bushuk (1966). Although it is likely that gluten still binds water after heating, this indicates that there may be great changes of the water distribution during heating. This, in turn, affects the reaction kinetics of the denaturation of protein and gelatinisation of remaining granular starch. Owing to these complex changes, the effects of heating will depend on the place in the bread. The DSC studies of gluten-starch model systems of Eliasson give some insights into this complexity (Eliasson, 1983; Fig. 16).

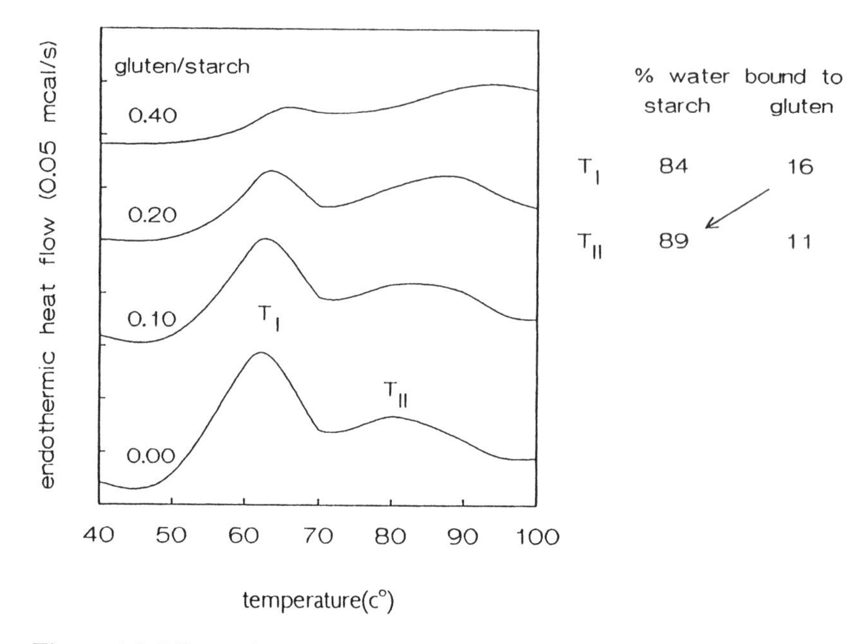

Figure 16. Effects of temperature and gluten content of gluten/starch mixtures (39-47% moisture) on the endothermic heat flow and transition temperatures, T_I and T_{II} and the calculated amounts of water bound to starch and gluten (Eliasson, 1983 ;reprinted with permission).

When the endothermic transitions of starch, containing 47% water, is determined by DSC, two transitions were observed: one around 60°C, T_I, and one around 80°C, T_{II}. This double transition indicates that starch was only partly gelatinised, owing to the limited amount of water. When gluten was added to the starch/water mixture, two changes may be observed. Starch gelatinised at a higher temperature (T_I and T_{II} increased) and the endotherm became smaller (Eliasson, 1983). The implication of the smaller endotherm is not clear. It may indicate that less energy is needed for gelatinisation, thus in a way compensating for the higher gelatinisation temperature, or that simply less starch is gelatinising. Similarly, in flour milling streams, the enthalpy decreased and the gelatinisation temparature of starch increased with increasing protein content (Eliasson et al, 1991).

Using the moisture dependence of the transition, it could be calculated that some, but not all, water bound to gluten became bound to starch. In addition, the amount of water bound to native gluten was only 16% (Eliasson, 1983), which was far lower than the amount of 31% found by Bushuk (1966). Although it is difficult to judge which value is correct, the difference is interesting since in the introduction it was demonstrated that the functional changes of gluten upon heating occur mainly between moisture contents of 20 to 25%. This indicates that there may be a very subtle and intricate balance between gluten denaturation and starch gelatinisation.

After heating the moisture-water activity relationship of gluten was not changed (Fig. 4; Weegels et al, 1994a). NMR studies have revealed that the interaction between water and heated gluten decreased and water was released to the bulk phase (LeGrys et al, 1981). These results may indicate that the binding strength of water to gluten proteins did not change after heating, but that the quantity of water bound has changed, probably owing to local aggregation phenomena discussed by Hermansson (1983ab). This results in a different `sponge' effect of gluten at a macromolecular level, but not of water binding on a molecular level.

In contrast, the interaction between water and starch after heating was strongly increased (Chinachoti et al, 1991). The increase in water binding to starch results in a decreased amount of water available for binding to other components and, in turn, affects the denaturation of protein and gelatinisation of starch. Owing to these complex changes, the effects of heating will depend on the place in the bread (Hoseney, 1984).

Research on factors that affect this balance is of high interest. In this respect, the work of Moore and Hoseney on dough expansion in an electrical resistance oven is important. Moore and Hoseney (1986) studied the influence of temperature on dough expansion in a electrical resistance oven. They found a lowering of the speed of expansion at 55°C and 80°C. The first change was attributed to a change of gluten and the second change to alterations of starch. Shortening and lipids interfered with the changes of volume increase. With shortening no deflection of the expansion rate at 55°C was observed. A similar observation was made when the flour was defatted, whereas polar lipids slowed down the expansion at 55°C more than normal (Fig. 17; Moore and Hoseney, 1986).

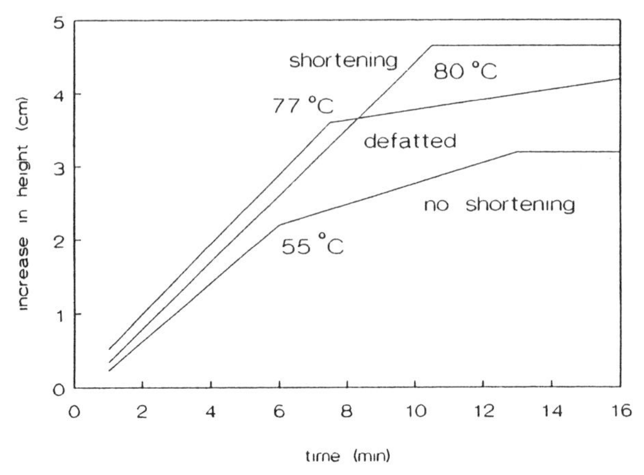

Figure 17. Effect of shortening added to flour and defatting of flour on the increase in dough height during heating of dough in an electrical resistance oven (Moore and Hoseney, 1986 ;reprinted with permission).

Pearce and coworkers (1987, 1988) found in ESR studies on the interaction between stearic acid and starch/gluten mixtures that stearic acid interacted mainly with starch. Upon heating the interactions were diminished. After cooling, stearic acid interacted more with gluten. These findings urge and challenge for more detailed investigations.

CONCLUSIONS

Despite the uniformity in dough size or batter density before baking, the major differences in product quality may become apparent during baking.

The reaction kinetics of the heat induced changes are governed mainly by time, temperature and moisture content. Loaf volume is a more sensitive parameter to determine the effects of heating than protein extractability.

During heating, great changes occur in gluten and starch. Depending on the characteristic under study, these changes occur over a broad range of temperatures. The temperatures, at which functional changes occur, do not always correspond to the temperatures, at which physico-chemical changes are observed. In addition, a delicate balance in water binding to starch and gluten exists that is of utmost importance to the gelatinisation and denaturation behaviour of these components.

Despite the importance of the heat induced changes for understanding and controlling product quality, a comprehensive study on the changes during baking is still missing. Therefore, research on the heat induced changes of dough remains an important and interesting challenge.

REFERENCES

ABLETT, S., BARNES, D.J., DAVIES, A.P., INGHAM, S.J., and PATIENT, D.W. 1988. ^{13}C and pulse nuclear magnetic resonance spectroscopy of wheat proteins. J. Cereal Sci. 7:11-20.
ALSBERG, C.L., and GRIFFING, E.P. 1927. The heat coagulation of gluten. Cereal Chem. 4:411-423.
ANNO, T. 1981. Studies on heat-induced aggregation of wheat gluten. J. Jap. Soc. Food Nutr. 34:127-132.
ARNTFIELD, S.D., and MURRAY, E.D. 1981. The influence of processing parameters on food protein functionality. I. Differential scanning calorimetry as an indicator of protein denaturation. Can. Inst. Food Sci. Technol. J. 14:289-294.
BAIANU, I.C., JOHNSON, L.F., and WADDELL, D.K. 1982. High-resolution proton, carbon-13 and nitrogen-15 nuclear magnetic resonance studies of wheat proteins at high magnetic fields: spectral assignments, changes with concentration and heating treatments of Flinor gliadins in solution - comparison with gluten spectra. J. Sci. Food Agric. 33:373-383.
BALE, R., and MULLER, H.G. 1970. Application of the statistical theory of rubber elasticity to the effect of heat on wheat gluten. J. Food Technol. 5:295-300.
BARR, P.J., and BARR, D.J. 1975. Improvements in pneumatic drying. Food Proc. Ind. 44:26;28, 31-32.
BECKER, H.A., and SALLANS, H.R. 1956. A study of the relation between time, temperature, moisture content, and loaf volume by the bromate formula in the heat treatment of wheat and flour. Cereal Chem. 33:254-265.
BELITZ, H.-D., KIEFFER, R., SEILMEIER, W., and WIESER, H. 1986. Structure and function of gluten proteins. Cereal Chem. 63:330-341.
BERLINER, E., and RÜTER R. 1928. Die Wirkung der Hitze auf Mehl. Z. Ges. Mühlenwesen 4:209-216.
BEVERIDGE, T., JONES, L., and TUNG, M.A. 1984. Progel and gel formation and reversibility of gelation of whey, soybean, and albumen protein gels. J. Agric. Food Chem. 32:307-313.
BLOKSMA, A.H., and BUSHUK, W. 1988. Rheology and chemistry of dough. Pages 182- 184 in: Wheat , Chemistry and Technology. Vol. II, Y. Pomeranz, ed. AACC: St. Paul.
BOOTH, M.R., BOTTOMLEY, R.C., ELLIS, J.R.S., MALLOCH, G., SCHOFIELD, J.D., and TIMMS, M.F. 1980. The effect of heat on gluten-physico-chemical properties. Ann. Technol. Agric. 29:399-408.
BUSHUK, W. 1966. Distribution of water in dough and bread. Bakers Dig. 40(10):38-40.
BUSHUK, W., and WINKLER, C.A. 1957. Sorption of water vapor on wheat flour, starch and gluten. Cereal Chem. 34:73-86.

CATSIMPOOLAS, N., and MEYER, E.W. 1970. Gelation phenomena of soybean globulins. I. Protein-protein interactions. Cereal Chem. 47:559-570.
CHINACHOTI, P., WHITE, V.A., LO, L., and STENGLE, T.R. 1991. Application of high- resolution carbon-13, oxygen-17, and sodium-23 nuclear magnetic resonance to study the influences of water, sucrose, and sodium chloride on starch gelatinization. 68:238-244.
CLUSKEY, J.E., and WU, Y.V. 1971. Optical rotary dispersion, circular dichroism, and infrared studies on wheat gluten proteins in various solvents. Cereal Chem. 48:203-211.
COCERO, A.M., and KOKINI, J.L. 1991.The study of the glass transition of glutenin using small amplitude oscillatory rheological measurements and differential scanning calorimetry. J. Rheol. 35:257-270.
COOK, W.H. 1931. Preparation and heat denaturation of the gluten proteins. Can. J. Res. 5:389-406.
COOKE, D., and GIDLEY, M.J. 1992. Loss of crystalline and molecular order during starch gelatinisation: origin of the enthalpic transition. Carbohydr. Res. 227:103-112.
DESHPANDE, S.S., and DAMODARAN, S. 1989. Heat-induced conformational changes in phaseolin and its relation to proteolysis. Biochim. Biophys. Acta 998:179-188.
DOGUCHI, M., and HLYNKA, I. 1967. Some rheological properties of crude gluten mixed in the Farinograph. Cereal Chem. 44:561-575.
DREESE, P.C., FAUBION, J.M., and HOSENEY, R.C. 1988a. Dynamic rheological properties of flour, gluten, and gluten-starch mixtures. I Temperature-dependent changes during heating. Cereal Chem. 65:348-353.
DREESE, P.C., FAUBION, J.M., and HOSENEY, R.C. 1988b. Dynamic rheological properties of flour, gluten, and gluten-starch doughs. II Effects of various processing and ingredient changes. Cereal Chem. 65:354-359.
ELIASSON, A.-C. 1983. Differential scanning calorimetry studies on wheat starch-gluten mixtures. J. Cereal Sci. 1:199-205.
ELIASSON, A.-C. 1987. Viscoelastic behaviour during the gelatinization of starch. II Effects of emulsifiers. J. Text. Stud. 17:357-375.
ELIASSON, A.-C., and HEGG, P.O. 1980. Thermal stability of wheat gluten. Cereal Chem. 57:436-437.
ELIASSON, A.-C., SILVERIO, J., and TJERNELD, E. 1991. Surface properties of wheat flour-milling streams and rheological and thermal properties after hydration. J. Cereal Sci. 13:27-31.
ELLIS, H.S., RING, S.G., and WHITTAM, M.A. 1989. A comparison of the viscous behaviour of wheat and maize starch pastes. J. Cereal Sci. 10:33-44.
EVERY, D. J. 1987. A simple, four-minute, protein-solubility test for heat damage in wheat. Cereal Sci. 6:225-236.
FEILLET, P., AIT-MOUH, O., KOBREHEL, K., and AUTRAN, J.-C. 1989. The role of low molecular weight glutenin proteins in the determination of cooking quality of pasta products: an overview. Cereal Chem. 66:26-30.

FIELD, J.M. TATHAM, A.S., and SHEWRY, P.R. 1987. The structure of a high-M_r subunit of durum-wheat *(Triticum durum)* gluten. Biochem. J. 247:215-221.
FOEGEDING, E.A. 1988. Thermally induced changes in muscle proteins. Food Techn. 42:58, 60-62, 64.
FUJIO, Y., and LIM, J.-K. 1989. Correlation between the glass-transition point and color change of heat-treated gluten. Cereal Chem. 66:268-270.
FUKUSHIMA, D. 1968. Internal structure of 7S and 11S globulin molecules in soybean proteins. Cereal Chem. 45:203-229.
GEDDES, W.F. 1929. Chemical and physico-chemical changes induced in wheat and wheat products by elevated temperatures - I. Can. J. Res. 1:528-557.
GEDDES, W.F. 1930. Chemical and physico-chemical changes induced in wheat and wheat products by elevated temperatures - II. Can. J. Res. 2:65-90.
GHIASI, K., HOSENEY, R.C., and VARRIANO-MARSTON, E. 1982. Effects of flour components and dough ingredients on starch gelatinization. Cereal Chem. 60:58-61.
GORDON, I.R., SMITH, N., GOUBAULT, F., and van SCHOUWENBURG, G. 1985. The present and future market for vital wheat gluten in the EEC. GIRA S.A. 1239 Collex: Geneva.
HERD, C.W. 1931. A study of some methods of examining flour, with special reference to the effects of heat. I. Effects of heat on flour proteins. Cereal Chem. 8:1-23.
HERMANSSON, A.-M. 1983a. Protein functionality and its relation to food microstructure. Qual. Plant. Plant Foods Hum. Nutr. 32:369-388.
HERMANSSON, A.-M. 1983b. Relationships between structure and waterbinding properties of protein gels. Pages 107-108 in: Research in Food Science and Nutrition. Vol. 2 Basic Studies in Food Science. J.V. McLoughlin, and B.M. McKenna, eds. Boole Press: Dublin.
HOOK, S.C.W. 1980. Dye-binding capacity as a sensitive index for the thermal denaturation of wheat protein. A test for heat-damaged wheat. J. Sci. Food Agric. 31:67-81.
HOSENEY, R.C. 1984. Starch and other polysaccharides; basic structure and function in food. Pages 27-45 in: Carbohydrates, Proteins, Lipids: Basic Views and New Approaches in Food Technology. F. Meuser, ed. Inst. Lebensmitteltechnologie, Getreidetechnologie: Berlin.
HOSENEY, R.C., ZELEZNAK, K., and LAI, C.S. 1986. Wheat gluten: a glassy polymer. Cereal Chem. 63:285-286.
HUEBNER, F.R., and BIETZ, J.A. 1993. Improved chromatographic separation and characterization of ethanol- soluble wheat proteins. Cereal Chem. 70:506-511.

HUTCHINSON, J.B., and BOOTH, R.G. 1946. The drying of wheat. IV. Phosphatase activity as an index of heat damage in cereals. J. Soc. Chem. Ind. 64:235-237.

JEANJEAN, M.F., DAMIDEAUX, R., and FEILLET, P. 1980. Effect of heat treatment on protein solubility and viscoelastic properties of wheat gluten. Cereal Chem. 57:325-331.

KALAB, M., and EMMONS, D.B. 1972. Heat-induced milk gels. V. Some chemical factors influencing the firmness. J. Dairy Sci. 55:1225-1231.

KALAB, M., VOISEY, P.W., HARWALKAR, V.R., and LAROSE, J.A.G. 1973. Heat- induced milk gels. VI. Effect of temperature on firmness in comparison with some common food gels. J. Dairy Sci. 56:998-1003.

KALICHEVSKI, M.T., and BLANSHARD, J.M.V. 1992. A study on the effect of water on the glass transition of 1:1 mixtures of amylopectin, casein and gluten using DSC and DTMA. Carbohydr. Polym. 19:271-278.

KALICHEVSKI, M.T., JAROSZKIEWICZ, E.M., and BLANSHARD, J.M.V. 1992a. Glass transition of gluten. I: Gluten and gluten-sugar mixtures. Int. J. Biol. Macromol. 14:257-266.

KALICHEVSKI, M.T., JAROSZKIEWICZ, E.M., and BLANSHARD, J.M.V. 1992b. Glass transition of gluten. 2: The effect of lipids and emulsifiers. Int. J. Biol. Macromol. 14:267-273.

KALICHEVSKI, M.T., JAROSZKIEWICZ, E.M., ABLETT, S., BLANSHARD, J.M.V., and LILLFORD, P.J. 1992c. The glass transition of amylopectin measured by DSC, DTMA and NMR. Carbohydr. Polym. 18:77-88.

KALICHEVSKI, M.T., JAROSZKIEWICZ, E.M., and BLANSHARD, J.M.V. 1993. A study of the glass transition of amylopectin-sugar mixtures. Polym. 34:346-358.

KASARDA, D.D., BERNARDIN, J.E., and GAFFIELD, W. 1968. Circular dichroism and optical rotatory dispersion of α-gliadin. Biochem. 7:3950-3957.

KENT-JONES, D.W. 1928. Some aspects of the effect of heat upon flour. Cereal Chem. 5:235-241.

KILARA, A. and SHARKASI, T.Y. 1986. Effects of temperature on food proteins and its implications on functional properties. CRC Crit. Rev. Food Sci. Nutr. 23:323-395.

KIM, C.S., and WALKER, C.E. 1992 Effects of sugars and emulsifiers on starch gelatinization evaluated by differential scanning calorimetry. Cereal Chem. 69:212-217.

LEGRYS, G.A., BOOTH, M.R., and AL-BAGHDADI, S.M. 1981. The physical properties of wheat proteins. Pages 243-264 in: Cereals: a renewable resource. Y. Pomeranz and L. Munck eds. AACC: St. Paul.

LEMESTE, M., HUANG, V.T., PANAMA, J., ANDERSON, G., and LENTZ, R. 1992. Glass transition of bread. Cereal Foods World 37:264-267.

LI-CHAN, E. 1983. Heat-induced changes in the proteins of whey protein concentrate. J. Food Sci. 48:47-56.
LIU, H., and LELIEVRE, J. 1993. A model for starch gelatinization linking differential scanning calorimetry and birefringe measurements. Carbohydr. Polym. 20:1-5.
LUPANO, C.E., and AÑON, M.C. 1986. Denaturation of wheat germ proteins during drying. Cereal Chem. 63:259-262.
LUSENA C.V., and ADAMS, G.A. 1950. Preparation of spray dried wheat gluten. Cereal Chem. 27:186-188.
MANGINO, M.E., HUFFMAN, L.M., and REGESTER, G.O. 1988. Changes in the hydrophobicity and functionality of whey during the processing of whey protein concentrates. J. Food Sci. 53:1684-1686, 1693.
MATSUDOMI, N., KATO, A., and KOBAYASHI, K. 1982. Conformation and surface properties of deamidated gluten. Agric. Biol. Chem. 46:1583-1586.
MCDERMOTT, E.E. 1971. The `turbidity' test as a measurement of thermal denaturation of proteins in wheat. J. Sci. Food Agric. 22:69-72.
MOORE, W.R. and HOSENEY, R.C. 1986. The effects of flour lipids on the expansion rate and volume of bread baked in a resistance oven. Cereal Chem. 63:172-174.
MOUNFIELD, J.D., HALTON, P., and SIMPSON, A.G. 1944. The drying of wheat. II. The drying of English wheat. J. Soc. Chem. Ind. 63:97.
MULLER, H.G. 1969. Application of the statistical theory of rubber elasticity to gluten and dough. Cereal Chem. 46:443-446.
NOEL, T.R., RING, S.G., and WHITTAM, M.A. 1990. Glass transitions in low-moisture foods. Trends Food Sci. Technol. 1 (3):62-67.
OKADA, K., NEGESHI, Y., and NAGAO, S. 1987. Factors affecting dough breakdown during overmixing. Cereal Chem. 64:428-434.
O'NEILL, T., and KINSELLA, J.E. 1988. Effect of heat treatment and modification on conformation and flavor binding by ß-lactoglobulin. J. Food Sci. 53:906-909.
PEARCE, L.E., DAVIS, E.A., GORDON, J., and MILLER, W.G. 1987. An electron spin resonance study of stearic acid interactions in model wheat starch and gluten systems. Food Microstruc. 6:121-126.
PEARCE, L.E., DAVIS, E.A., GORDON, J., and MILLER, W.G. 1988. Electron spin resonance studies of isolated gluten systems. Cereal Chem. 65:55-58.
PENCE, J.W., MOHAMMAD, A., and MECHAM, D.K. 1953. Heat denaturation of gluten. Cereal Chem. 30:115-126.
POMERANZ, Y. 1988. Thermolabilität und Thermostabilität von Prolaminproteinen - Beziehungen zum Backverhalten. Getr. Mehl Brot 42:355-357-357.

PRIVALOV, P.L., and KHENCHINASHVILI, N.N. 1974. A thermodynamic approach to the problem of stabilization of globular protein structure: a calorimetric study. J. Mol. Biol. 86:665-684.
RIZVI, S.S.H., JOSEPHSON, R.V., BLAISDELL, J.L., and HARPER, W.J. 1980. Separation of soy-spun fiber, egg albumen, and wheat gluten blend by sodium dodecyl sulfate gel electrophoresis. J. Food Sci. 45:958-961.
RUDZIK, L., WÜST, E., LAVEN, F.-J., and MARQUARDT, U. 1987. Deconvolution am Beispiel von Infrarotspektren verschiedener Milcheiweißstoffe. Z. Lebensm. Unters. Forsch. 185:271-274.
RUSSELL, P.L. 1987. Gelatinisation of starches of different amylose/amylopectin content. A study by differential scanning calorimetry. J. Cereal Sci. 6:133-145.
SARKKI, M.-L. 1979. Food uses of wheat gluten. J. Amer. Oil Chem. soc. 56:443-446.
SCHOFIELD, J.D., BOTTOMLEY, R.C., TIMMS, M.F., and BOOTH, M.R. 1983. The effect of heat on wheat gluten and the involvement of sulphydryl-disulphide interchange reactions. J. Cereal Sci. 1:241-253.
SCHOFIELD, J.D., BOTTOMLEY, R.C., LEGRYS, G.A., TIMMS, M.F., and BOOTH, M.R. 1984. Effects of heat on wheat gluten. Pages 81-90 in: Gluten Proteins, Proceedings of the 2nd International Workshop on Gluten Proteins. A. Graveland and J.H.E. Moonen, eds. PUDOC: Wageningen.
SCHOFIELD, J.D., and BOOTH, M.R. 1983. Wheat proteins and their technological significance. Pages 1-65 in: Developments in food proteins - 2. B.J.F. Hudson, ed. Applied Science Publishers: London.
SCHREIBER, H., MÜHLBAUER, W., WASSERMAN, L., and KUPPINGER H. 1981. Reaktionskinetische Untersuchungen über den Einfluß der Trocknung auf die Qualität von Weizen. Z. Lebensm. Unters. Forsch. 173:169-175.
SHEWRY, P.R., HALFORD, N.G., and TATHAM, A.S. 1992. High molecular weight subunits of wheat glutenin. J. Cereal Sci. 15:105-120.
SLADE, L., LEVINE, H., IEVOLELLA, J., and WANG, M. 1993. The glassy state phenomenon in applications for the food industry: application of the food polymer science approach to structure-function relationships of sucrose in cookie and cracker systems. J. Sci. Food Agric. 63:133-176.
STENVERT, N.L., MOSS, R. and MURRAY, 1981. The role of dry vital wheat gluten in breadmaking. Part I. Quality assessment and mixer interaction. Bakers Dig. 55(4):6-7,10,12.
STEWART II, W.E, de SOMER, P., and de CLERCQ, E. 1974. Protective effects of anionic detergents on interferons: reversible denaturation. Biochim. Biophys. Acta 359(1974)364-368.
SUGIYAMA, T., RAFALSI, A., PETERSON, D., and SÖLL, D. 1985. A wheat HMW glutenin subunit gene reveals a highly repeated structure. Nucl. Acids Res. 13:8729-8737.

TAKEDA, K., SASA, K., NAGAO, M., and BATRA, P.B. 1988. Biochim. Biophys. Acta 957:340-344.
TANFORD, C. 1980. The Hydrophobic Effect: formation of Micelles and Biological Membranes. Wiley-Interscience, John Wiley & Sons: New York.
TATHAM, A.S., FIELD, J.M., SMITH, S.J., and SHEWRY, P.R. 1987. The conformations of wheat gluten proteins, II, aggregated gliadins and low molecular weight subunits of glutenin. J. Cereal Sci. 5:203-214.
TESTER, R.F., and MORRISSON, W.R. 1990. Swelling and gelatinization of cereal starches. I. Effects of amylopectin, amylose and lipid. Cereal Chem. 67:551-557.
UMBACH, S.L., DAVIS, E.A., GORDON, J., and CALLAGHAN, P.T. 1992. Water self- diffusion coefficients and dielectric properties determined for starch-gluten-water mixtures by microwave and by conventional methods. Cereal Chem. 69:637-642.
VOUTSINAS, L.P., NAKAI, S., and HARWALKAR, V.R. 1983. Relationships between protein hydrophobicity and thermal functional properties. Can. Inst. Food Sci. Technol. J. 16:185-190.
de VRIES, U., SLUIMER, P., and BLOKSMA, A.H. 1988. A quantative model for heat transport in dough and crumb during baking. Pages 174-188 in: Cereal Science and Technology in Sweden. Proc. Int. Symp. June 13-16 1988: Ystad.
WALL, J.S., and BECKWITH, A.C. 1969. Relationship between structure & rheological properties of gluten proteins. Cereal Sci. Today 14(1):16-18, 20-21.
WEEGELS, P.L. 1991. Sensitive measurement of contents and reactivities of thiol and disulphide in soluble and insoluble proteins. Pages 414-419 in: Gluten proteins 1990. Proceedings of the 4th International Workshop on Gluten Proteins. W. Bushuk and R. Tkachuk, eds. AACC: St. Paul.
WEEGELS, P.L. 1994. Depolymerisation and re-polymerisation of wheat glutenin during dough processing and effects of low M_r wheat proteins. PhD Thesis, King's College, University of London: London.
WEEGELS, P.L., and HAMER, R.J. 1989. Predicting the baking quality of gluten. Cereal Foods World 34:210-212.
WEEGELS, P.L., VERHOEK, J.A., de GROOT, A.M.G., and HAMER, R.J. 1994a. Effects on gluten of heating at different moisture contents. I. Changes in functional properties. J. Cereal Sci. 19:31-38.
WEEGELS, P.L., de GROOT, A.M.G., VERHOEK, J.A., and HAMER, R.J. 1994b. Effects on gluten of heating at different moisture contents. II. Changes in physico-chemical properties and secondary structure. J. Cereal Sci. 19:39-47.
WEIPERT, D., and GERSTENKORN P. 1988. Beurteilung der Klebereigenschaften mittels der Kriecherholungsmessung. Getr. Mehl Brot 42:99-103.

WESTERLUND, E., THEANDER, O., and ÅMAN, P. 1989. Effects of baking on protein and aqueous ethanol- extractable carbohydrate in white bread fractions. J. Cereal Sci. 10:139-147.
WICKER, L., LANIER, T.C., HAMANN, D.D., and AKAHANE, T. 1986. Thermal transitions in myosin-ANS fluorescence and gel rigidity. J. Food Sci. 51:1540-1543, 1562.
WRIGLEY, C.W.. du CROS, D.L., ARCHER, M.J. DOWNIE, P.G., and ROXBURGH, C.M. 1980. The sulfur content of wheat endosperm proteins and its relevance to grain quality. Aust. J. Plant Physiol. 7:755-766

CHAPTER 6
LIPIDS, LIPID-PROTEIN INTERACTIONS AND THE QUALITY OF BAKED CEREAL PRODUCTS

Didier Marion[@],Laurence Dubreil[@],
Peter J.Wilde[£] and David C.Clark[£]
(@) I.N.R.A. Laboratoire de Biochimie et Technologie des Protéines BP1627
44316 Nantes cédex 03 (France) and
(£) Institute of Food Research, Norwich laboratory, Norwich Research Park,
Norwich NR4 7UA (United Kingdom).

INTRODUCTION

Lipids are either ingredients -i.e. fat and emulsifiers - using in the formulation of baked cereal products or components of wheat flour (about 2% of dry wheat flour). Whatever their content and their origin, lipids are essential in providing products with good organoleptic properties. Lipid functionality is expressed through interactions with other components of dough including water. Especially, lipid-protein interactions have attracted attention of the cereal scientists since more than 40 years. A sudden awareness for their importance in dough formation has emerged probably with the discovery that most wheat flour lipids are recovered in extracted wheat gluten and has been strengthened by the fact that wheat flour lipids become unextractable by apolar solvents such as petroleum ether or hexane in dough and gluten (Olcott and Mecham, 1947). These observations have stimulated the idea that wheat lipids are essential for the formation of gluten networks and expression of their viscoelastic properties. Different models associating gliadins-glutenins and lipids or gliadin-glutenin and starch have been proposed to explain the role of lipid-protein interactions in the dough and gluten viscoelasticity. The model developed by Hess and Mahl in 1954 where lipoproteins form a sticky layer around starch granules, is the best illustration of the early interest in interactions between lipids and proteins and their role in dough elasticity. This model has been followed by many others as the lipoprotein model of Grosskreutz (1961), Hoseney et al., (1970), and Wherli and Pomeranz (1970).

However, most of these models were drawn from defatting and reconstitution experiments or from complex fractionation procedures of wheat proteins using aqueous and organic solvents. These experiments are quite delicate to interpret because the organisation of wheat doughs is greatly modified and also because isolation of lipid-rich protein fractions does not necessary mean that interactions occur between these components. Using non-perturbing spectroscopic techniques and freeze-fracture electron microscopy, it has been proved that no lipoprotein complexes are formed between gluten proteins and lipids and that lipid are not essential to the gluten formation (Marion et al., 1987; Hargreaves et al., 1995).

To explain the role of lipids through their interactions with proteins, two physicochemical events are essential: (1) the oxido-reducing mechanism involving lipoxygenase catalysed oxidation of polyunsaturated fatty acids and rearrangement of protein disulphide bonds of gluten proteins (2) the mechanism which involves lipids and lipid-protein complexes in the formation and stability of air-water (foam) and oil-water (emulsion) interfaces during dough mixing, proofing and baking. The former way has been previously reviewed (Frazier, 1983; Nicolas and Drapron, 1983) and any important progress has been realized recently on this topic. In this chapter, we will focus on the latter mechanism for which new data obtained recently offer promising perspectives of improving wheat quality through breeding programs and genetic engineering as well as for the improvement of the quality of baked cereal products through the use of well adapted lipid ingredients and lipolytic enzymes.

STRUCTURE AND ULTRASTRUCTURE OF LIPIDS FROM GRAIN TO BAKED PRODUCTS: A CHALLENGE FOR THE STUDY OF LIPID-PROTEIN INTERACTIONS

Chemical Diversity and Heterogeneous Distribution of Lipids Through the Wheat Kernel

Lipids are a family of small molecules which exhibits a great structural diversity. In a first attempt more than 1000 different molecular species are present in a biological material like wheat seed when combinations of head groups and fatty acids are considered. Furthermore, many other lipophilic components are solubilized in acyl lipids (e.g. sterols, carotenoids, tocopherol) (see Barnes, 1983; Morrison, 1988 for reviews).

This biodiversity - to use an up to date concept - is a problem for the analysis of lipids in biological samples such as wheat grain or wheat flour. The most complete data on the composition of wheat acyl lipids has been determined by Morrison and co-workers using different thin layer chromatographic procedures for the fractionation of lipid classes and gas chromatography for the analysis of fatty acids (Hargin and Morrison, 1980) (Figure 1). Since that work, no more data have been obtained on the composition of wheat lipids and the most important progress has been realized with the development of high performance liquid chromatography techniques allowing a rapid characterisation of wheat lipids and their molecular species (Marion et al., 1984; Christie and Morrison, 1988). The structural diversity of lipids is complicated by their heterogeneous distribution in cells and tissues. Thus, in aleurone layer and embryo, more than 70% of the total lipids are composed by neutral lipids (mainly triglycerides). On the contrary more than 70% of the starchy endosperm total lipids are composed by polar lipids, phospholipids and glycolipids. Furthermore, some lipids of the starchy endosperm which are tightly bound into the starch granules exhibit a quite different composition from the non starch lipids. Starch lipids are mainly composed of monoacyl lipids, LPC and FFA (see Figure 1) which could form complexes with the amylose helix.

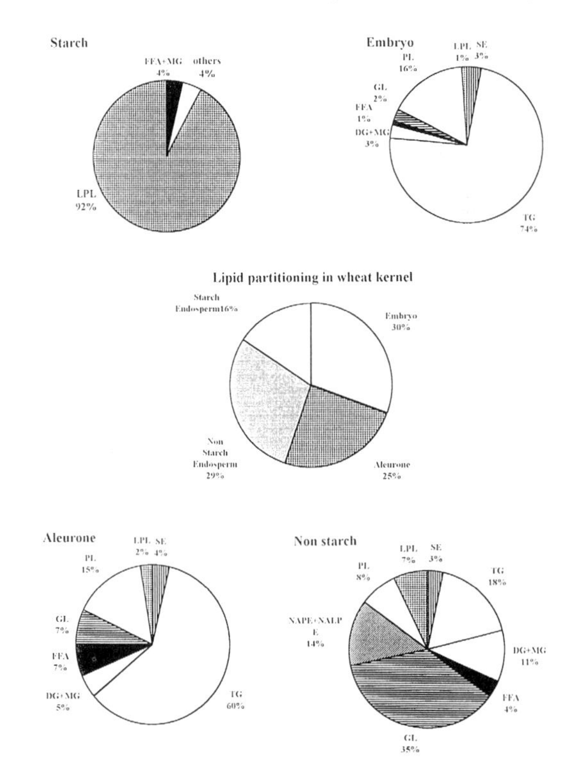

Figure 1. Composition of wheat kernel lipids. SE: sterol esters; TG, triglycerides; MG, monoglycerides; DG, diglycerides; FFA, free fatty acids; GL, glycolipids are composed mainly by monogalactosyldiglycerides (MGDG) and digalactosyldiglycerides (DGDG); minor GL are monogalactosylmono-glycerides (MGMG) and digalactosylmonoglycerides (DGMG); PL, phospholipids are composed mainly of PE, phosphatidylethanolamine (PE); N-acylphosphatidylethanolamine (NAPE), N-acyllysophosphatidylethanolmaine (NALPE), phosphatidylcholine (PC), phosphatidylinositol (PI); LPL, lysophospholipids are mainly composed of lysophosphatidylcholine (LPC); minor PL and PL are phosphatidylglycerol (PG), diphosphatidylglycerol or cardiolipin (DPG), phosphatidic acid (PA), lysophosphatidylethanolamine (LPE); lysophosphatidylinositol (LPI). Adapted from Hargin and Morrison (1980)

The composition of polar lipids differs between aleurone layer, embryo and starchy endosperm. For example, starchy endosperm are unique among living organisms since its phospholipid fraction is composed of more than 50% of NAPE/NALPE, which are not found in aleurone layer and embryo (Hargin and Morrison, 1980)(Figure 1). Finally, lipids are not randomly distributed in the different tissues, but are involved in the formation of cell and organelle membranes which imposes both lipid-lipid and lipid-protein interactions. For example, the triglycerides present in embryo and aleurone cells are stored in oil bodies, an oil droplet surrounded by a phospholipid-protein monolayer (Murphy, 1990) while endoplasmic reticulum, precursor of protein bodies, is mainly composed of phospholipids. Glycolipids, DGDG and MGDG, are the main components of the amyloplast membranes (Fischwick and Wright, 1980). On the contrary the fatty acid composition of wheat acyl lipids is relatively homogeneous with 50-60% of linoleic acid (18:2 n-6), 15-20% of palmitic acid (16:0), 8-20% of oleic acid (18:1 n-9) and only 3-5% of linolenic acid (18:3 n-3) and 1-3% of stearic acid (18:0). However, starch lipids diverge significantly from this fatty acid composition with a higher content in palmitic acid (Morrison, 1988).

The first important consequence of this heterogeneous distribution of lipids is that a variation in the size and quantity of oil bodies, amyloplasts and cells leads to changes of lipid content and composition of wheat grain. Secondly, it is obvious from this spatial distribution of lipids in wheat kernel, that milling process should influence the composition as well as the lipid content of wheat flour. For example, increasing wheat flour yield should increase contamination with embryo and aleurone tissues and therefore, the triglyceride content of wheat flour. In this regard, it has been shown that about 30% of apolar wheat flour lipids are germ oil (Stevens, 1959).

The structural diversity as well as the tissular and subcellular distribution of lipids constitutes a challenge for studies on the role of lipids in wheat technology. This means that reconstitution experiments which necessary involve extraction and reinsertion of lipid fractions have to be interpreted cautiously since such manipulations lead to a new system diverging from the organisation of the components of native flour. This has been largely ignored by most cereal scientists and have precluded them for a long time to carry out research works on the organisation of lipids in wheat grain, flour and doughs.

Water Dependent Rearrangements of Lipids from Wheat Grain to Dough

Lipid self-assembly

Owing to their hydrophobicity, lipids self-assemble in aqueous solvents to give rise to different types of situations: some lipids are completely insoluble and form crystals or oil phases, while others, due to their amphiphilic structure, swells to form liquid-crystalline (L.C.) phases or dissolve to form micelles (Small, 1986). Micelles is the most simple and stable self-assembly formed by lipids above a critical concentration and temperature. Below their critical micellar concentration these lipids exist as monomers in true solution.

According to the structure of the lipid molecule (Table I), discoidal, spherical and rod-shaped micelles can be formed (Small, 1986). This phase is adopted only by monoacyl lipids as sodium salts of free fatty acids and lysophospholipids. Polar lipids can also swell in lamellar, hexagonal or cubic L.C. phases. The structure of these L.C. phases has been extensively investigated since the pioneering work of Luzzati and co-workers (Luzzati and Husson, 1962; Luzzati et al., 1968). The lamellar phase is formed by stacking of lipid bilayers separated by water lamellae which give generally rise, in excess of water, to multi layered or unilamellar vesicles called liposomes (Table I). Biomembranes are composed of a single lipid bilayer (Singer and Nicolson, 1972). Two type of hexagonal phases are found: the normal or H_I type hexagonal phase and the reverse or H_{II}type hexagonal phase.

TABLE I

Polymorphic phases, molecular shapes and the critical packing parameter for the major wheat lipids. Adapted from Gennis (1986) and Faucon and Meleard (1993).

POLAR LIPIDS	PHASE	MOLECULAR SHAPE	$v/a_0.l$
lysoPL FFA (sodium salts)	micellar hexagonal HI	inverted cone	<1/3 (sphere) 1/3-1/2 (cylinder,rod)
PC, PI, PG PA, PS PE (saturated) DGDG APE (saturated) ALPE	lamellar (bilayer)	cylinder	1/2-1
PE (unsaturated) APE (unsaturated) PA + Ca2+ APE+Ca2+ MGDG	hexagonal HII	cone	>1

In the H_I type, lipids are packed in long cylinders with the polar head group exposed at the surface; such phase are generally formed at high concentrations of lipids which dissolve in micelles in excess of water. In the H_{II} type, lipid polar head groups face inside the cylinders and form a water channel (Seddon, 1987) (Table I). The cubic phase is certainly the most complex L.C. structure for which structural data are relatively recent. It is generally adopted in some physicochemical conditions (pH, concentration, temperature.) by nonbilayer lipids, i.e. lipids forming micelles or H_{II} L.C. phases, (see Lindblom and Rilfors, 1989 for a review). The liquid-crystalline polymorphism of lipids is closely related to their structure (structure of the polar head group and of the hydrophobic tail) and to the physicochemical characteristics of the medium (pH, temperature, ionic strength, water content.). In excess of water and at room temperature, some lipids preferentially adopt bilayer phases such as PC and DGDG, while this phase is formed in specific conditions for others such as MGDG, PE and NAPE. For these lipids, liquid-crystalline phase transitions can occur on heating, pH changes or presence of divalent cations. For example the mono- unsaturated dioleyl-PE exhibits a bilayer phase below 10°C while a transition to an inverted H_{II} phase is observed on heating above this critical temperature (Cullis and De Kruiff, 1978). N-acylation of PE leads to an important increase of this transition temperature since it is necessary to heat an aqueous liposomal suspension of N-oleyl-PE to more than 50°C to complete the lamellar to hexagonal transition (Akoka et al., 1988). This bilayer stabilizing effect of N-acylation has been attributed to an increase of head group hydration and to the formation of intermolecular hydrogen bonds between amide group (Lafrance et al. 1990). A quite simple geometric approach has been defined to explain the presence of the different phases and a relation has been found between the surface of the polar head group (a_0), the volume of the hydrophobic tail (v) and the length of the hydrophobic tail (l). For different values of the ratio v/a_0l, called the critical packing parameter, a lipid molecule can be assumed to a geometric figure, for example as a cone, an inverted cone or a cylinder (Cullis and De Kruijff, 1979; Gennis, 1986; Faucon and Meleard, 1993) which lead to specific lipid aggregates in aqueous solutions (Table I). With this concept most of the L.C. phase properties of lipids can be explained and predicted. The most interesting consequence of this concept is that an equimolar mixture of lipids having normal cone and inverted cone shapes will give rise to a cylinder shape and therefore exhibits a lamellar structure in water. This has been confirmed experimentally for example in the case of mixture of unsaturated PE and LPC (Madden and Cullis, 1982). Exhaustive reviews have been done on the polymorphic behaviour of amphiphiles (Cullis and De Kruijff, 1979; Cullis et al, 1986; Seddon, 1987; Marsh, 1991; Tate et al, 1991).

Phase Behaviour and Liquid-crystalline Polymorphism of Wheat Lipids: the First Step in the Expression of Wheat Lipid Functionality.

The phase behaviour of lipids can become very complicated in systems where more than a single lipid molecule is concerned. This has been emphasized with the work of Larsson and co-workers (Carlson et al., 1978; Carlson et al., 1979; Larsson, 1986) on the water-dependent polymorphic phase behaviour of wheat lipids (Figure 2). These phase diagrams showed for the first time that lipid-lipid and lipid-water interactions could play a major role in cereal processing.

They also highlight the weakness of reconstitution experiments as a unique approach in the determination of the functional role of lipids. However, the main drawback of these phase diagrams is that they have been constructed from extracted wheat lipids so that the natural lipid associations have been lost.Therefore, it was not possible to predict the behaviour of wheat lipids in situ.

Freeze-fracture electron microscopy appeared as the most relevant method to probe lipid polymorphism in flours and doughs. This technique has provided new exciting and unexpected results on the liquid-crystalline organisation of wheat polar lipids (Al-Saleh et al., 1986; Marion et al., 1987; Marion et al., 1989). In wheat endosperm of mature dry seeds lipids form aggregates mainly of the nonbilayer types forming long tubules or granular structures characteristic of L.C. hexagonal II and cubic phases, respectively. When water diffuses in the dry endosperm numerous vesicles are observed growing from the non lamellar aggregates. This transition is also observed during hydration of wheat flour and it appears that the cubic phase could be an intermediate step in this hexagonal to lamellar phase transition. Such L.C. transitions have already been observed in model phospholipid systems and in this case, the vesicles obtained are of the unilamellar type (Vail and Stollery, 1979). In wheat doughs numerous oil droplets are observed which correspond certainly to the oil bodies found in aleurone layer and embryo. Therefore, at the end of dough mixing and in extracted wheat gluten only oil droplets and lamellar vesicles are present (Marion et al., 1987; Marion et al., 1989). These transitions lead to a better dispersion of lipids and especially of polar lipids into the dough. It is interesting to note that the observation of such lipid phase transitions from flour to dough and gluten is in agreement with the phase diagram obtained for extracted wheat polar lipids (Carlson et al., 1978). Below 15% water content only hexagonal phases are observed while between 15 and 50% only lamellar structures are predicted (Figure 2). However the L2 phase described by these authors has not been observed.

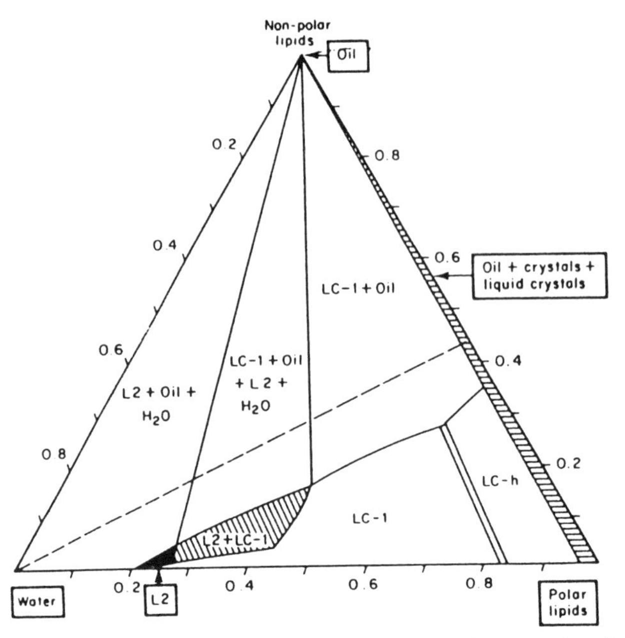

Figure 2. Phase equilibria in the ternary system of non polar lipids/polar lipids/water. Lipids were extracted from wheat flour with water saturated n-butanol. LC-l, liquid-crystalline lamellar, LC-h, liquid-crystalline hexagonal; L2, inverse micellar LC phase (cubic). (from Carlson et al., 1978)

Firstly, the presence of hexagonal phases in dry endosperm means that a transition from lamellar biomembranes to hexagonal phases should occur during the final dehydration step of grain maturation (Figure 3). Such a transition has been already observed in biological membranes (Crowe and Crowe, 1982; Gordon-Kamm and Steponkus, 1984). Secondly, since these lipid structures derive from biological membranes, lipids are probably associated with specific hydrophobic membrane proteins. These membrane vesicles and oil bodies have been isolated by mild extraction procedures and their observation by freeze-fracture electron microscopy revealed some particles on their surface (Marion et al., 1989). These particles are typical features of transmembrane proteins according to freeze-fracture studies carried out on biomembranes. The proteins present in these vesicles are composed of proteins with polypeptide molecular mass ranging from 15 to more than 90 kDa as shown by sodium dodecyl sulfate polyacrylamide gel electrophoresis (Marion et al., 1989). The most abundant proteins have molecular mass around 15 kDa. It has been shown that the lamellar to hexagonal phase transition causes lateral segregation of lipids and membrane proteins and thereby expulsion of the so-formed protein aggregates (Simon 1978; Gordon-Kamm and Steponkus, 1984). In the case of wheat, the fate of these membrane proteins is still unknown but different possibilities could be reasonably considered: aggregation to gluten proteins, reincorporation into lipid vesicles and oil droplets, formation of oligomeric globulin-like proteins (Figure 3).

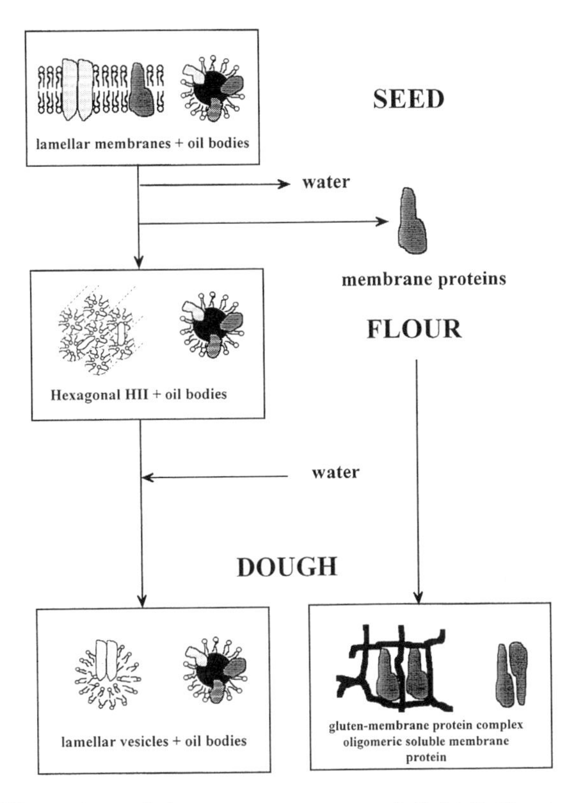

Figure 3. Illustration of the rearrangements of lipid L.C. structures and lipoproteins from seed to dough deduced from freeze-fracture electron microscopy data (Al-Saleh et al., 1986; Marion et al., 1987; Marion et al., 1989).

Therefore, it is highly probable that some membrane proteins are not recovered in the lipid fraction of dough. Membrane proteins are largely ignored in cereal technology while about 10% wheat proteins could have a membrane origin if we consider that the mean lipid/protein weight ratio in membranes is 1:1 and there is about 1% membrane lipids in a wheat flour. This deficiency should be filled in the future since cell biology has provided new techniques to isolate and characterize such proteins (Gennis, 1986).

Finally, the hexagonal to lamellar L.C. phase transition occurring on wheat flour hydration can explain the difference in the extractibility of wheat lipids with non polar organic solvents when wheat flour is hydrated and mixed into dough. The hexagonal H_{II} phase for which the fatty backbone of polar lipids are exposed at the surface of this structure should be easily accessible to such solvents contrarily to bilayers where the fatty chains are protected by the polar head group interfaces. Therefore, it is thus possible that the hexane or petroleum ether extractable lipids, also called "free lipids", are probably more markers of the hexagonal/bilayer proportion in wheat flours than of lipid-protein interactions as it is generally assumed.

THE TRANSFER OF POLAR LIPIDS FROM THE AQUEOUS PHASE OF DOUGH TO AIR-WATER AND OIL WATER INTERFACES: ROLE OF LIPID-LIPID AND LIPID-WATER INTERACTIONS.

A close relationship has been found between the effect of wheat lipids on the bread volume and their effect on the foaming properties of the aqueous phase of dough. Especially, it has been shown that non polar lipids, triglycerides and free fatty acids which are anti-foaming agents are also detrimental to bread volume. On the contrary polar lipids that improve the foaming properties of dough liquor are also good bread improvers.

These experiments carried out by MacRitchie and co-workers (MacRitchie and Gras, 1973; MacRitchie, 1983) revealed that the formation of a dough foam and its stability during resting, proofing and baking is fundamental to ensure the formation of aerated bread crumb. Furthermore, these experiments revealed that the aqueous phase of dough plays a major role in the formation of a dough foam. In cake batters emulsification by proteins and polar lipids ensures a good dispersion of fat and could contribute to liquid-solid phase separation into the batter (Le Roux et al., 1990). The transfer of the lipid vesicles from the bulk water to the air-water or oil-water interfaces is therefore a key mechanism for the expression of lipid functionality.

Two types of films are formed when bilayer liposomes spread at air-water interfaces. At zero surface pressure, there is a slow transformation of the closed bilayer into an open monolayer at the air-water interface. When bilayer liposomes are spread against a surface pressure, a part of the multi-lamellar structure is preserved. Furthermore, it has been shown that the outer layer spread at a higher yield than the inner layer (Pattus et al., 1978). Lipid exchange occurs between the interfacial monolayer, the immediate under layer of vesicles in interaction with the monolayer and the vesicles in bulk solution (Schindler, 1979). This equilibrium is controlled by the lipid-lipid interactions in the bilayer which produce an energy barrier for the bilayer-monolayer transformation.

For example, the strong lipid-lipid interactions in bilayer liposomes composed of saturated PC prevent the spontaneous adsorption of these phospholipids at the air-water interface (Figure 4) (Notter, 1984). When defects are introduced in lipid packing, adsorption can occur. These defects can be created by nonbilayer lipids such as lipid forming micelles (LPC for example) or hexagonal II L.C. phases (unsaturated PE for example) (Figure 4).

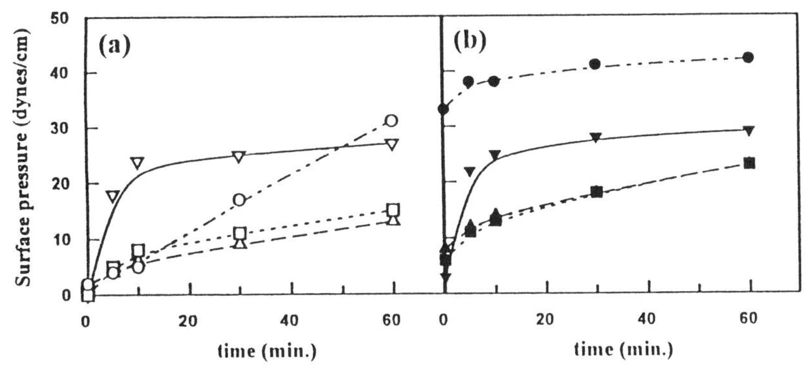

Figure 4. Adsorption of phospholipid mixtures dispersed as (a) multi-lamellar (vortexing) and (b) small unilamellar (sonicated) liposomes. (▽), DPPC:LPC 7:3 w/w; (△)DPPC:egg PG 7:3; (□)(■) DPPC:egg PE 75:25 ; (○)(●), DPPC:soy PE 75:25 . (adapted from Notter, 1984).

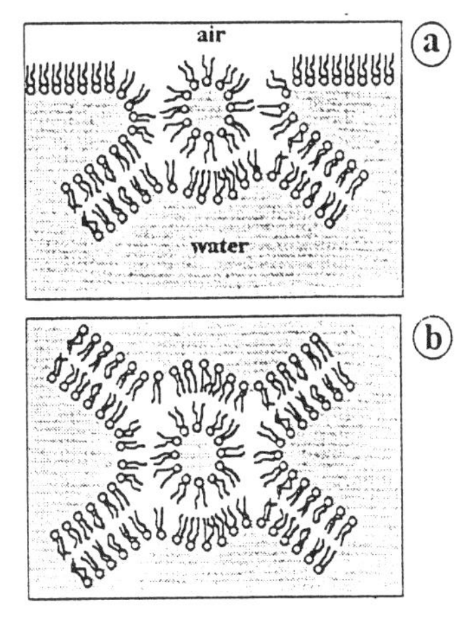

Figure 5. Inverted micelles in bilayers as possible common structure formed on membrane fusion and spreading of bilayer liposomes at air-water interface. Adapted from Cullis and De Kruijff (1978).

A mechanism similar to this occurring on bilayer fusion would proceed at the air-water interface (Pattus et al., 1978; Yu et al., 1984). During fusion it has been shown that an inverted micellar structure is formed by the nonbilayer lipids at the contact between bilayers. A similar mechanism could occur at the contact between bilayer and the air-water interface which would subsequently stimulate the spreading of lipids (Figure 5). Defects can also occur if the curvature of the bilayer is important. This is the case of small unilamellar vesicles which spread more rapidly at air-water interfaces than larger multilayered liposomes (Figure 4b). In this regard, it is noteworthy that (1) wheat lipids are composed of nonbilayer lipids such as MGDG or NAPE (Hargin and Morrison, 1980) and (2) small bilayer vesicles with a relatively important curvature are formed on wheat flour hydration and dough mixing (Marion et al., 1987; Marion et al., 1989). In this regard, it has been shown that the positive effect of lecithins on bread volumes is more important when these phospholipids are added as small sonicated unilamellar vesicles than as large multilayered lecithin liposomes (Rajapaksa et al., 1983).

PROTEINS AND THE EXPRESSION OF THE FUNCTIONALITY OF LIPIDS

Gluten proteins do not form lipoprotein complexes with lipids as it has been shown by combining freeze-fracture electron microscopy and phosphorus NMR (Marion et al., 1987). However, phosphorus NMR has shown that the viscoelasticity influences the dynamics of vesicles dispersed in gluten network (Marion et al., 1987). On the contrary, lipids do not play a role on the gluten viscoelasticity (Hargreaves et al., 1995). Therefore, it is highly probable that the viscoelastic dough network controls only the expansion of gas bubbles stabilized by lipoprotein films during proofing and baking. This view is supported by breadmaking experiments carried out by adding lipids to defatted wheat flour of good and poor qualities. The loaf volume-lipid content curves exhibit quite similar shapes with only translation towards higher volumes or higher lipid content (MacRitchie, 1983).

In a dough system, the proteins which are susceptible to play a direct role in the stabilization of air-water interfaces are those present in the aqueous phase of dough, in which are also dispersed the lipid vesicles (Marion et al., 1987). Therefore, albumins, globulins and lipid binding proteins which are with polar lipids, the other surface active components of wheat flour present in the aqueous phase of dough should contribute to the spreading and stability of lipid and lipid-protein films.

Wheat Lipid-binding Proteins: A Brief Survey of their Structural Diversity.

A lipid binding protein is a protein which is naturally associated to lipid aggregates or which is able to bind spontaneously lipids or lipid aggregates without supply of energy. This definition means that only nondenatured proteins have to be considered and therefore excludes proteins which emulsify oil and fat due to an interfacial denaturation under physical treatment (mixing for example). The search of lipid binding proteins in wheat flour began quite early with the isolation of a lipoprotein complex in the petroleum ether extract from wheat flour (Balls and Hale, 1940). Thereafter, many other fractionation procedures have been used to extract lipid binding proteins from wheat flour using both aqueous or organic solvents. These procedures lead to fractions containing both proteins and lipids but this mixed systems are not a proof that lipoprotein complexes do exist. Different proteins have been isolated and characterized as thionins (Balls and Hale, 1940), chloroform-methanol extracted (CM)-proteins (Redman and Ewart, 1973), ligolin (Frazier et al., 1981) and S-proteins (Zawistowska et al., 1985; Zawistowska et al., 1986). Lipid binding has been proved only in the case of thionins and ligolins (Bekes and Smied, 1981; Frazier et al., 1981).

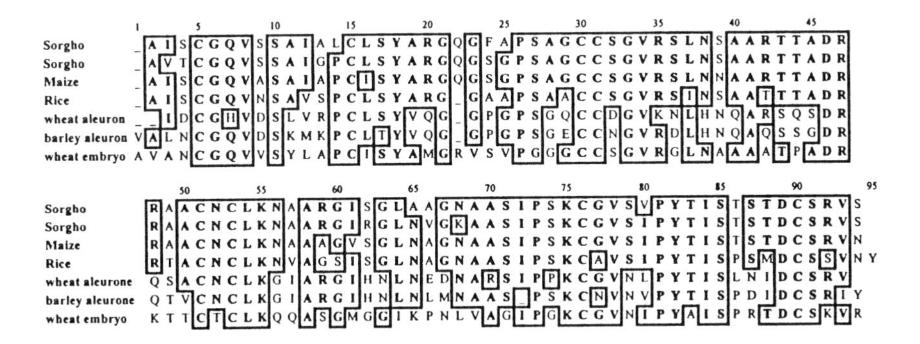

Figure 6. Amino acid sequences of wheat aleurone (Desormeaux et al., 1992), wheat embryo (Neumann et al., 1994), sorghum (Pelese-Siebenbourg et al., 1994), maize (Tchang et al., 1988) and barley (Skriver et al., 1992) 9kDa LTPs.

Nevertheless, most of the protein isolated until now shared some structural features as low molecular weight, high cystine and basic amino acid contents. In a first attempt, it was obvious that such characteristics are also shared by plant lipid transfer proteins (LTP). LTPs are abundant and ubiquitous plant proteins which are characterized by a molecular weight of about 9KDa, a basic pI and 8 cysteines forming 4 disulphide bonds (Kader, 1993). These proteins are non specific since they are capable to catalyse the intermembrane transfer of different types of polar lipids and to bind free fatty acids and acylco-enzymeA (Kader, 1993). An homologous protein which accounts for about 2% of wheat soluble proteins has been purified from wheat endosperm and its sequence has been determined (Desormeaux et al., 1992) (Figure 6).

Another isoform of LTP has been found in wheat embryo which is susceptible to be phosphorylated by a wheat calcium dependent kinase (Neumann et al., 1993). The presence of different isoforms suggest that these wheat LTPs belong to a multigenic family. This is in agreement with previous observations done for castor bean where the 4 LTP isoforms found are organ-specific (Tsuboi et al., 1991). The wheat endosperm LTP is localised in the aleurone cells (Dubreil et al., 1994) in agreement with the barley LTP (Skriver et al., 1992). This peripheral localisation explain why only 50% of the aleurone LTP is recovered in wheat flour (Desormeaux et al., 1992). Therefore, the milling process could influence both the lipid (see §1.1) and the lipid binding protein content of wheat flour.

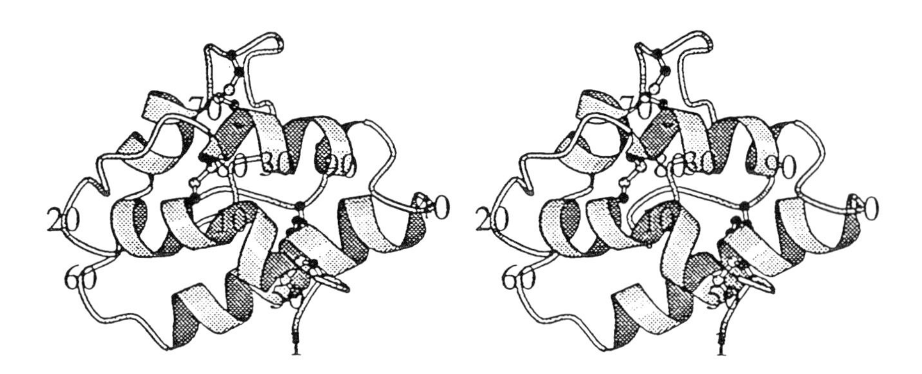

Figure 7. Stereo view of the three dimensional structure of wheat LTP. (from Gincel et al., 1994)

The structure of the wheat lipid transfer protein has been recently determined from multidimensional NMR data (Simorre et al., 1991; Gincel et al., 1994). The polypeptide backbone is composed of a bundle of 4 helices linked by flexible loops which is packed against a N-terminal fragment having a non standard saxophon shape (Figure 7). An hydrophobic tunnel formed by residues located in the second half part of the protein is the lipid binding site (Gomar et al., 1996). This is a new and original type of structure for a lipid binding protein and has been subsequently confirmed for barley and maize LTPs, both by NMR and X-ray crystallography (Sinh et al., 1995; Gomar et al., 1996; Heinemann et al., 1996). The transfer mechanism involves both adsorption of the protein at the bilayer interface and binding of a lipid molecule. Lipid binding induces a significant increase of the overall helicity of the protein (Desormeaux et al., 1992) which facilitates protein crystallization (Pebay-Peyroula et al., 1992). When disulfide bonds are broken the helix content decreases and lipid transfer is suppressed (Desormeaux et al., 1992; Kader, 1993). These results suggest that the helical structure is essential in the formation of the hydrophobic tunnel (Gincel et al., 1994). The adsorption site to the membrane interface is still unknown but orientational studies on lipid monolayer by ATR spectroscopy and fluorescence microscopy show that adsorption is accompanied by non negligible rearrangements of both lipids and protein (Subirade et al., 1995). Furthermore, it has been shown that lipid transfer activity could be improved by increasing the content of anionic phospholipids in liposomes (Petit et al., 1994). This improvement could be related to the distribution of charged residues and the subsequent electrostatic potential repartition at the protein surface (Gomar et al., 1996). However, the interaction with lipid bilayers is weak and it is impossible to isolate a stable liposome-LTP complex (Desormeaux et al., 1992; Subirade et al., 1993). This weak interaction is quite in agreement with the intermembrane exchange-transfer of lipids catalysed by LTP.

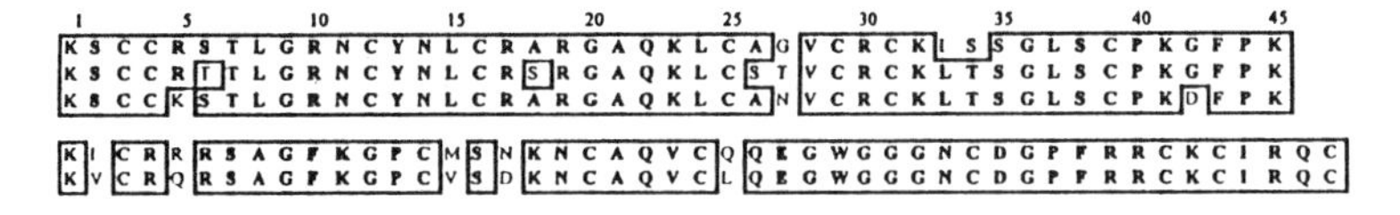

Figure 8. Amino acid sequences of α, β and γ-purothionins (Mak and Jones, 1976; Jones and Mak, 1977; Colilla et al., 1990).

Thionins, the first lipid binding proteins isolated in wheat seeds (Balls and Hale, 1940) were also the first for which interactions with lipid components was proved *in vitro* (Hernandez-Lucas et al; 1977; Bekes and Smied, 1981) as well as in vivo (Carbonero et al., 1980). Two different types of proteins are known: the α-b type and the g-type (Mak and Jones, 1976; Jones and Mak, 1977; Colilla et al., 1990) (Figure 8).

The latter has been discovered recently and differ from the former by its sequence and folding pattern (Bruix et al., 1993). The three-dimensional structure of a-b thionins has the shape of a Greek letter G with the vertical stem composed of two antiparallel helices and extended strands in the horizontal arm (Clore et al., 1987; Teeter et al., 1990) . In contrast, g-thionin have only one helix parallel to a three stranded b-sheet (Bruix et al., 1993). Their lipid binding properties are stronger than those of LTP since it is possible to isolate a thionine-membrane and thionin-lipid complexes (Bekes and Smied, 1981; Carbonero et al., 1980). This suggests that thionins are able to penetrate deeply in membranes in agreement with their effect on the permeability of bacterial and fungal membranes (Carrasco et al., 1981).

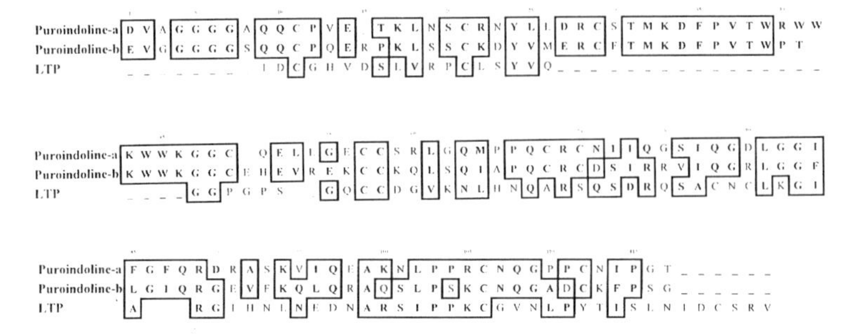

Figure 9. Amino acid sequence alignment of puroindolines (Blochet et al., 1993; Gautier et al., 1994) and wheat LTP (Desormeaux et al., 1992).

Such membranotoxic properties are also found for protein toxins from animal venoms which are generally cystine-rich and basic low molecular weight proteins. It is interesting to note that the three-dimensional structure of g-thionins is quite similar to the folding pattern of toxic proteins isolated from venoms of scorpions (Bruix et al., 1993). In a sense, these membranotoxic or membrane active proteins behave as transmembrane proteins. Therefore, we have used a procedure which is specific to isolate hydrophobic transmembrane proteins in cells: Triton X114 phase partitioning (Bordier, 1981; Pryde and Phillips, 1986). Nonionic detergents due to their amphiphilic structure are able to compete and replace the natural membrane lipids and due to their detergent properties to solubilize the protein in mixed protein-detergent micelles. TX114 is especially interesting since above 25°C aggregation of micelles takes place so that two phase are formed and separated after centrifugation: an upper detergent depleted phase and a lower detergent-rich phase. Transmembrane proteins are found in the TX114 rich phase and in the case of wheat flour, it has been shown that non membrane proteins are also found. In regard to their partitioning behaviour, they have probably the ability to penetrate bilayer membranes. This hypothesis has been strenghtened by the fact that thionins, known for their membranotoxic effect were among the isolated proteins (Blochet et al., 1991). Surprisingly, the major protein found in this phase was a new basic and cystine rich low molecular weight protein. This protein was purified and its sequence determined (Blochet et al., 1993). It contains 10 cysteines forming 5 disulphide bridges and exhibits an unique tryptophan-rich domain (Trp-Arg-Trp-Trp-Lys-Trp-Trp-Lys) which has led to call this protein, puroindoline (from the Greek word puros for wheat and indoline for the indole ring of tryptophan). Another minor isoform of this protein has been isolated (Blochet et al., 1991) and subsequent cDNA sequencing (Gautier et al., 1994) revealed that it has a truncated tryptophan rich domain (Trp-Pro-Thr-Lys-Trp-Trp-Lys). Therefore, the former isoform was named puroindoline-a and the latter puroindoline-b. Puroindolines do not exhibit sequence homology with any other known cystine-rich wheat protein but it is possible to find a relatively good alignment of the sequences of LTP and puroindolines except in a zone containing the tryptophan-rich domain. (Figure 9). This suggests that LTP and puroindoline tridimensional structures are closely related. Finally, the structural homology suggests that the tryptophan-rich domain could be responsible for the transmembrane protein-like properties of puroindolines. In this regard, it should be noted that tryptophans are frequently found near the lipid/water interface for membrane spanning proteins and other lipid binding proteins (Weiss et al., 1991; Schiffer et al., 1992; Landolt-Marticorena et al., 1993).

Interestingly, it has been shown that a close relationship exist between puroindolines and friabilins (Jolly et al. 1993; Morris et al., 1994), proteins found at the surface of starch granules (Greenwell and Schofield, 1986). Since these proteins are only found on starches from soft wheats, Greenwell and Schofield (1986) have suggested that they could play a role on the endosperm texture. Thereafter, it has been shown that only a minor part of these proteins are found on the surface of starch granules (Jolly et al., 1993) probably because they complex lipids found on granule surface during the extraction of starch (Greenblatt et al., 1994). The localisation of puroindolines in the aleurone layer confirmed that these basic proteins can only be a genetic and not a functional marker of grain softness/hardness (Dubreil et al., 1994).

The relationships between LTPs and puroindoline with previously described lipid binding proteins are interesting, especially in the case of ligoline. Ligoline has a molecular weight of about 9kDa and its amino acid composition reveals some characteristics shared by wheat LTP: absence of methionine residues, the presence of two tyrosines and about the same amount of basic amino acids, arginine and lysine. However in contrast to ligoline, LTP does not contain phenylalanine and other significant differences appear in the amount of leucine, glutamic, threonine and histidine residues (Frazier et al., 1981) (Table II).

The lower cysteine content of ligoline could be attributed mainly to chemical degradation during hydrolysis in HCl 6N. Therefore, these analogies suggest that ligoline is a mixture of low molecular weight proteins in which LTP could be the major protein. However, the main difference between LTP and ligoline concerns their lipid binding properties. It has been suggested that ligoline binds triglycerides in a molar ratio close to 1:1. while

TABLE II

Comparison of the amino acid composition (weight %) of wheat LTP (Desormeaux et al., 1992), puroindolines (Blochet et al., 1993;Gautier et al., 1994), ligoline (Frazier et al., 1981) and CM proteins (Garcia-Maroto et al., 1990).

Amino Acid	Proteins						
	Puroindoline-a	Puroindoline-b	CM1	CM3	CM16	LTP	Ligoline
ASX	9.4	4.3	9.1	5.4	6.6	15.5	8.2
GLX	11.9	14.6	8.3	11.0	15.1	6.6	11.8
SER	4.3	6.9	7.5	6.9	5.8	10	11.8
GLY	13.6	12	9.1	6.9	5.8	10	9.4
THR	3.4	2.6	5.0	4.1	7.5	0.9	4.7
ALA	3.4	2.6	5.8	6.2	4.2	4.4	8.2
PRO	7.7	6	8.3	10.4	9.2	6.6	5.9
TYR	0.8	0.8	5.0	4.1	4.2	2.2	2.4
VAL	3.4	5.2	6.6	6.9	4.2	6.6	5.9
MET	1.7	2.6	1.6	1.3	4.2	-	-
CYS	8.5	8.6	8.3	6.9	8.4	8.8	3.5
ILE	5.9	3.4	4.1	4.1	4.2	6.6	3.5
LEU	6.8	5.2	6.6	10.4	8.4	7.7	8.2
PHE	2.5	4.3	1.6	2.7	1.6	-	4.7
TRP	4.3	3.4	-	2.0	1.6	-	nd
HIS	-	0.8	2.5	1.3	0.8	3.3	1.2
ARG	6.8	7.7	7.5	5.5	5.8	6.6	5.9
LYS	5.1	8.6	2.5	2.0	1.6	3.3	4.7

LTP is unable to bind such non polar lipids (Kader, 1993). However, it is important to notice that triglyceride binding was assayed using triolein radiolabelled with ^{14}C on the fatty acid moiety so that it is not clear if this protein has bound an intact triglyceride or a monoacylated derivatives (monooleylglycerol or oleate) since lipases are present in wheat flour (Galliard, 1983). Concerning the CM and S proteins which are obviously the same proteins (Zawistowska et al., 1985), it is interesting to note that they are not recovered in the detergent-rich phase but in the upper detergent depleted phase suggesting that they are not able to bind strongly lipids in a normal aqueous system (Blochet et al., 1991).

Many other lipid binding proteins are present in the wheat soluble protein fraction and especially in the TX114 extract (Blochet et al., 1991). Some proteins have been forgotten as the metal-dependent phospholipid-binding proteins (Fullington, 1967), which are probably homologous to animal and plant annexins (Kee, 1988; Smallwood et al., 1990). These proteins, for which the real biological activity is still unknown, exhibit a quite interesting inhibitory activity of phospholipase A2 (Kee, 1988). As mentioned above (see § 1.2.2), membrane spanning proteins have been ignored - probably owing to their great structural diversity and to the lack of efficient extraction and purification procedures - while they can represent an important part of wheat proteins on a quantitative point of view. In a first attempt, the oil bodies proteins (Murphy, 1990) which form a lipoprotein monolayer at the surface of oil droplets should be considered since they are good candidates in the stabilization of oil-water interfaces in doughs.

Non Lipid Binding Soluble Proteins and Polar Lipids compete for the Interfaces

The majority of soluble proteins have a high affinity for the interface, which they saturate at much lower concentrations than low molecular weight surfactants (Chen et al., 1993). This is consistent with the ability of proteins to cause a greater lowering of the interfacial tension on a mole for mole basis at low concentrations. However, at higher concentrations the converse is true, since a pure surfactant stabilised interface generally has a lower interfacial tension than that formed from adsorbed protein. Thus, the relative amounts of protein and lipid present in solution can influence the composition of the interfacial layer in simple non-interacting mixtures of protein and lipid. In solutions containing low concentrations of lipid, the protein will dominate the adsorbed layer; conversely protein will be displaced from the interface in solutions containing high lipid concentrations.

However, the situation is often more complex due to interactions between the two components. Here an additional component in the form of the lipid/protein complex must be considered. This component may possess significantly different properties compared to free lipid and protein. Studies of these complex systems has been mostly restricted to high hydrophobic lipophilic balance (HLB) surfactant + protein systems, due to the low solubility of lipids (Dickinson and Woskett, 1989). An example of the changes observed in surface tension in a system where interactions between components occur is shown in Figure 10. The system illustrated is comprised of β-lactoglobulin, the whey protein from milk and a micellar phospholipid, palmitoyl-LPC. Two data curves are presented showing the surface tension properties of the surfactant in the presence and absence of the protein (Sarker et al., 1995). The features described above are evident with the protein dominating the interfacial tension properties at low concentrations and the surfactant dominating at higher concentrations. Interaction between the components is revealed by the cross over of the curves. This occurs because the free surfactant concentration is lowered by the amount complexed with the protein. In this example, the binding process is characterised by a dissociation constant (Kd) of 166μM (Sarker et al., 1995). We can imagine a similar mechanism with lipid transfer proteins which are able to bind lysophospholipids more strongly (dissociation constant Kd in the micromolar range) than b-lactoglobulin (Gomar et al., 1996). In this regard, it should be noted that the homologous protein from barley is a major component of beer foam (Sorensen et al., 1993).

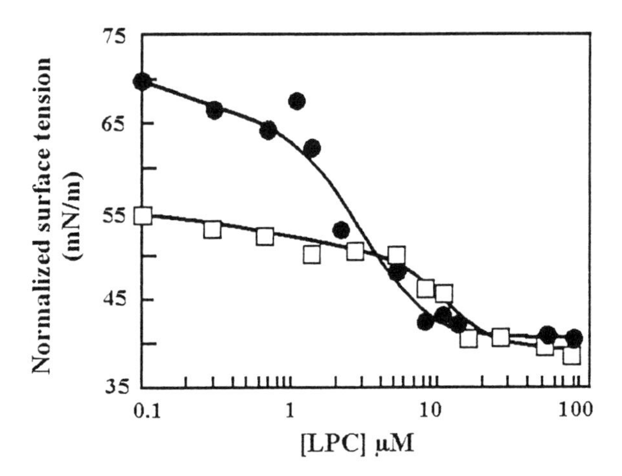

Figure 10. The surface tension of palmitoyl-LPC, as a function of increasing concentration in the absence (●) and presence (□) of 0.1mg/ml β-lactoglobulin. From Sarker et al., (1996).

The composition of many foods results in the adsorbed layers at the interfaces of food foams and emulsions often containing both protein and lipids. The stability of such dispersions is very complex and is often only observed if appropriate temperature conditioning steps are taken to ensure that the lipid globules achieve the correct solid/liquid ratio following crystallisation of triglycerides. This is very necessary in the case of dairy foams, such as whipped cream. In this system, air bubbles are initially stabilised by the adsorption of soluble milk proteins. It is only after extended whipping that fat globules accumulate at the air-water interfaces, start to bridge between air bubbles and contribute significantly to the structure and texture of the product (Brooker, 1993). Adsorption and spreading of liquid fat at the interfaces of foam lamellae (thin films) can induce film rupture and bubble coalescence in some products. This is particularly a problem in cases where there is high dispersed phase volume (e.g. in a reasonably drained foam (Coke et al., 1990) or creamed emulsion) or the system is exposed to further processing involving high shear forces (Chen et al., 1993).

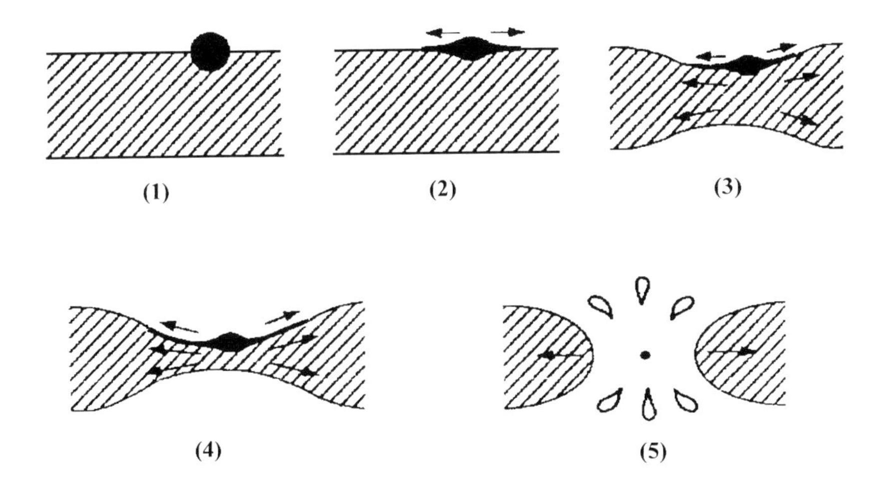

Figure 11. A schematic representation of the different steps in the spreading of a lipid droplet (black spot) causing local film thinning leading to film rupture. From Marion and Clark (1996)

The adsorption of molten fat droplets and the subsequent spreading of lipid causes thin film rupture by a Marangoni effect (Figure 11). Some interlamellar liquid in the thin film is associated with the polar head groups of the lipid and is dragged away from the point of adsorption of the lipid droplet by the spreading lipid. This causes local thinning of the thin film and increases the probability of film rupture.

Disruption of the structure of the adsorbed interfacial layer can occur as a result of adventitious adsorption of lipid monomers or small aggregates such as micelles. This can also cause instability in the dispersion. This arises from the different stabilisation mechanisms displayed by proteins and lipids (Figure 12). In the case of lipids, provided the sample is above the transition temperature, the lipid molecules adsorbed at the interface are capable of lateral diffusion in the plane of the adsorbed layer (Lalchev et al., 1994). Thus, if the interface is expanded (dilation), causing a localised increase in interfacial tension, adsorbed lipid molecules can diffuse laterally from regions of lower interfacial tension (Figure 12a). This process acts to restore equilibrium interfacial tension. In contrast, protein molecules in the adsorbed layer interact with each other to form a elastic (Clark et al., 1993), immobile (Clark et al., 1990) adsorbed layer rather like a rubber sheet (Figure 12b). It is easy to imagine how such a structure acts to dissipate interfacial expansion over a large area of interface, in the manner in which the rubber skin of a balloon stretches as it is inflated.

However, such a mechanism of stabilization is only effective whilst interactions between neighbouring molecules are maintained. If interactions are weak or have been destroyed in certain regions, expansion of the interface results in failure or tearing in the weak region. This is precisely what happens in an interfacial layer comprised of both protein and lipid (Figure 12c). In such a mixed system, both molecules compete for interfacial area.

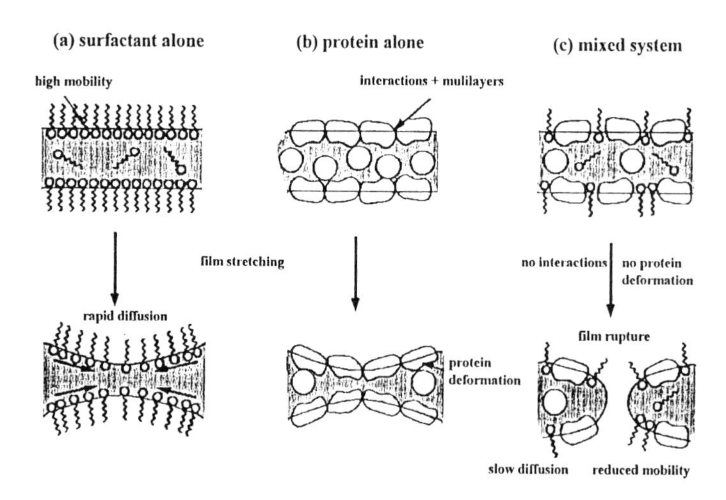

Figure 12. Schematic diagram showing the different mechanisms of thin film stabilization. (a) The Marangoni mechanism in surfactant films; (b) Instability in mixed component films; (c) The viscoelastic mechanism in protein-stabilized films. From Marion and Clark (1996).

The importance of foam lamellae (thin film) stability in baking has not been studied systematically. One limiting factor relates to the technical problem of working with insoluble gluten. However, it is likely that the stability of these structures plays an important role during the preparation and baking of loaves and sponge cakes. The quality of these foods is inseparably linked to the expansion of gas bubbles during the proofing and baking stages. Scanning electron micrographs reveal that some bubbles grow to such an extent that they are separated from neighbours only by thin films, the aqueous interlamellar phase of which is essentially devoid of the gluten matrix (Gan et al., 1990). Therefore, prior to starch gelatinisation and the formation of a solid sponge during the cooking process, bubbles in these foods are most probably stabilised by thin films stabilised by adsorbed layers of soluble cereal proteins and lipids.

Lipid-binding Proteins on Lipid Spreading and Film Stability at Air-water Interfaces

Proteins can stimulate Adsorption and Spreading of Lipids at Air-water Interfaces

As previously discussed the formation of inverted micelles in bilayers could facilitate the spreading of lipids at air-water interfaces. Inverted micelles can be formed when peptides and proteins interact with the lipid bilayer interface (Van Echteld et al., 1982; King et al., 1984; Killian et al., 1996). Especially, proteins or peptides which exhibit a high affinity for nonbilayer lipids induce a lateral segregation of these lipids and a local formation of nonbilayer structures which can favour spreading. Such mechanisms have been described in the case of pulmonary surfactant proteins. Pulmonary surfactant is a mixture of phospholipids and proteins, which helps the lungs expand by lowering the surface tension at the air/liquid interface in the alveoli (Hawgood and Clements, 1990). The main phospholipid component of lung surfactant is DPPC which has the ability to greatly lower the surface tension. However, DPPC does not exhibit rapid adsorption and spreading. Other unsaturated phospholipids (PI, PE, PG) (Notter, 1984) and especially, specific proteins contribute to increasing greatly the spreading kinetics (Hawgood et al., 1987)(Figure 13).

The most efficient lung surfactant proteins are low molecular weight amphipathic proteins. For example, SP-B is an amphiphilic basic and cystine-rich protein of 79 residues, structural features which remind those of wheat lipid binding proteins (Hawgood and Clements, 1990).

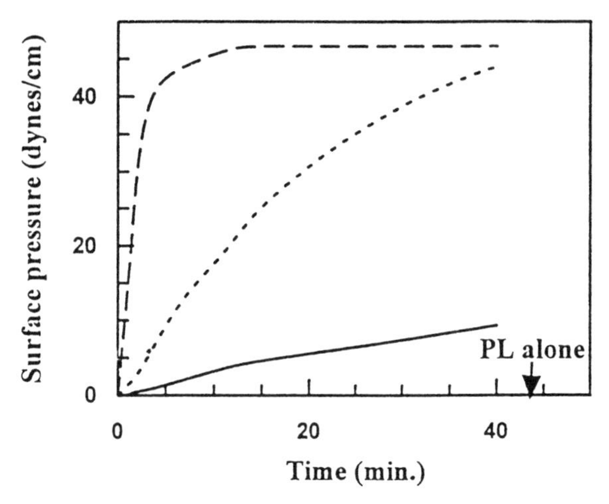

Figure 13. Stimulation of phospholipid surface film formation by pulmonary surfactant proteins (SP). Adsorption .kinetics of a phospholipid mixture DPPC-PG 4:1 alone (arrow); (——) phospholipid mixture + SP28-36 (28-36kDa); (-------), phospholipid mixture + SP5-18 (5-18kDa); (— — —), phospholipid mixture + SP5-18 + SP28-36. From Hawgood et al., (1987).

LTPs can facilitate the exchange of lipids between the monolayer and the underlayer of liposomes, and such a mechanism has been postulated to explain the spreading of lipids as a monolayer (Pattus et al., 1978). Furthermore, these proteins, which act only on the outer bilayer lipid leaflets may facilitate in some cases the transbilayer movement of phospholipids from the inner to the outer membrane leaflet (Wirtz et al., 1986). This would serve to increase the yield of lipid transferred from the inner bilayer to the monolayer at the air-water interface (Schindler, 1979). Pressure-area isotherm of dipalmitoylphosphatidylglycerol (DPPG) and DPPG-wheat LTP monolayers showed that LTP penetrates the phospholipid film and binds DPPG at low surface pressure (10-15mN/m). At 20mN/m, LTP is transferred under the film to be finally expulsed in the subphase as a DPPG-LTP complex at 28mN/m (Figure 14). These results suggest that the wheat LTP can catalyse the exchange-transfer of polar lipids between bilayer liposomes dispersed in the aqueous phase and the monolayer at air-water interface. Such a mechanism involving wheat LTP is highly dependent on the surface pressure of the lipid monolayer (Subirade et al., 1995).

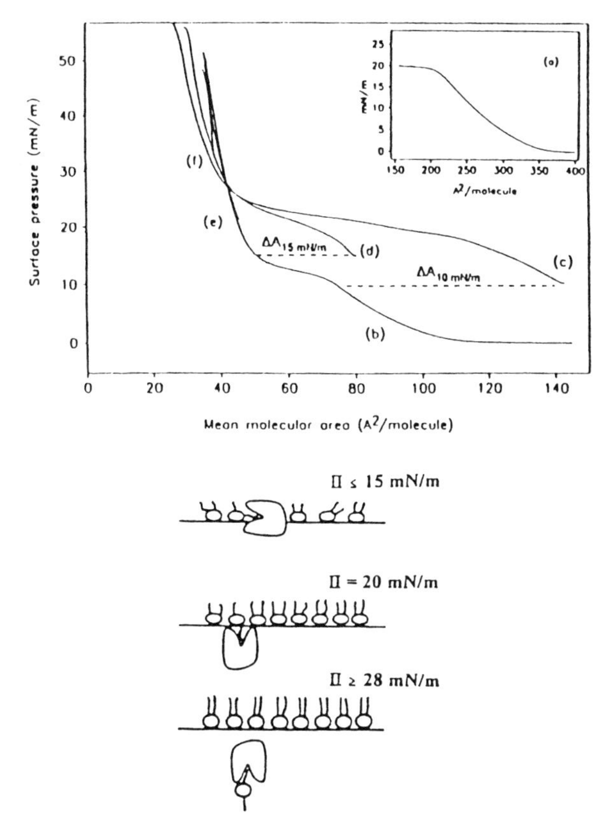

Figure 14. (1) Compression isotherms for LTP alone (a) and for DPPG monolayer in the absence (b) and presence of wheat LTP injected at 10 (c), 15 (d), 20 (e) and (f) 35mN/m; (2) illustration of the different steps involved in the LTP/DPPG interactions deduced from the study of the lipoprotein film by infrared spectroscopy and fluorescence microscopy. From Subirade et al. (1995).

The Puroindoline-lipid Films: An Example of a Synergistic Stabilisation

The functional properties of puroindoline-a have recently been investigated in the presence of palmitoyl-LPC (Wilde et al., 1993). Puroindoline-a alone has excellent properties significantly superior to commonly studied globular proteins. Unexpectedly, the foam stability of puroindoline-a increased markedly in the presence of added LPC and this enhancement was significantly greater than that expected from the sum of the individual properties. This synergistic effect was maximal between molar ratios (R) of 1 and 10 moles of LPC per mole of puroindoline-a (Figure 15).

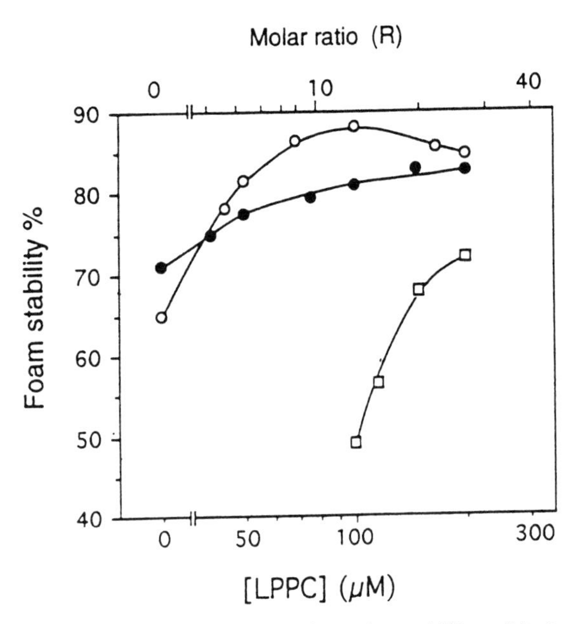

Figure 15. The effect of palmitoyl-LPC on the stability of (○), puroindoline-a and (●), puroindoline-b foams; (□), concentration dependence of the stability of foams formed from palmitoyl-LPC alone. From Husband et al., (1995).

The mechanism responsible for the enhancement of foaming properties is still unclear. However, it is evident that the enhancement is observed under conditions where both puroindoline and LPC are present in the interfacial layer (Wilde et al., 1993). In addition, the protein is known to bind the lipid analogue Puroindoline-b which exhibits similar foaming properties to puroindoline-a but lower affinities for polar lipids (Kd for palmitoyl-LPC are 55μM and 1.04mM for puroindoline-a and puroindoline-b, respectively) does not lead to a similar foam enhancement (Husband et al., 1995) (Figure 15). It seems likely that the complex formed has enhanced surface properties and this could have important technological significance as it has been observed in some model food systems. For example, preliminary results have revealed that low levels of puroindoline-a that comprise only 1% (10-20μg/ml) of the total protein load present in beer, can restore the foaming properties to beer adulterated with stearic acid, phospholipid or triglyceride (Clark et al., 1994). Similar results have been obtained concerning the foaming properties of egg white proteins adulterated with oil (Husband et al., 1995) (Figure 16). Since a nonionic detergent is necessary to extract puroindolines (Blochet et al., 1991; Blochet et al., 1993), it is highly probable that such lipid-puroindoline complexes are present in bread dough and may play an important role during proofing and baking.

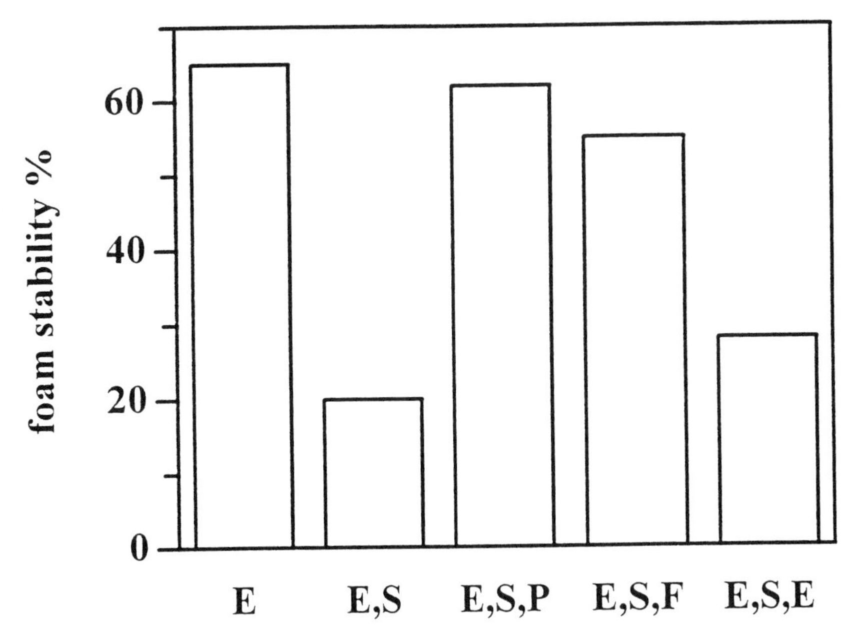

Figure 16. The effect of puroindolines on egg protein foams adulterated by oil. (E) egg-white protein (1mg/ml) alone and in the presence of (E,S), 4.3% soya oil; (E,S,P), 4.3% soya oil and 0.1mg/ml puroindoline-a; (E,S,F) 4.3% soya oil and 0.1mg/ml puroindoline-b; (E,S,E) 4.3% soya oil and 0.1mg/ml egg white protein. From Husband et al., (1995).

TOWARDS THE ENGINEERING OF LIPIDS AND LIPID-PROTEIN INTERACTIONS FOR IMPROVING WHEAT QUALITY

In this brief review we have confronted the data obtained on the physicochemical properties of lipids and lipid-protein interactions in model systems with what it is known on the effect of lipids in breadmaking technology. Most of the relationships that we have suggested here are still untested but they allow us to propose some new strategy for improving the wheat quality through modification of lipid functionality. The simple fact that surfactants improve bread crumb texture means that the polar lipid content and composition of wheat are not optimal. In the future, the challenge will be to find the means of improving the surface properties of wheat polar lipids and to limit the detrimental impact of non polar lipids. The genetic route is less easy for lipids than for proteins, since many enzymes and therefore many genes are involved in the synthesis of lipids. We have shown in this review that introducing some nonbilayer lipids is quite favourable to the transformation of closed bilayer liposomes dispersed in the aqueous phase of dough to an open monolayer at air-water interfaces. This can be done either by choosing specific surfactants and surfactant mixtures, or by modifying chemical structure of lipids through use of enzymes and especially hydrolytic enzymes. In this regard lipases are good candidates because they are able to generate polar lipids - monoglycerides- from apolar triglycerides and detergent-like molecules from phospholipids and glycolipids (lysophospholipids, galactosylmonoglycerides). However, the main drawback in the use of such enzymes is that they also generate free fatty acids. These lipid components are known to be deleterious to the quality of cereal products so that it is necessary to limit their presence in wheat doughs.

Finally, the best way is certainly to improve the functionality of lipids through lipid binding proteins. For example, proteins such as puroindolines can act synergistically with polar lipids to improve the stability of lipoprotein films or to prevent the destabilization of protein foams by non polar lipids. Since these proteins are encoded by a single or a limited number of genes, it is possible to introduce these proteins in breeding programs. Furthermore, the transgenic approach offers fascinating opportunities for manipulation of the genes coding for such proteins in order to improve their expression, change their localisation and their functionality using directed mutagenesis. Increasing the content of these proteins could also be a way to improve the effect of commonly used bread surfactants. As shown in the case of beer foam, this could also avoid the negative effect of free fatty acids generated by the use of lipases.

REFERENCES

AKOKA S., TELLIER C., LE ROUX C., MARION D. (1988) A phosphorus magnetic resonance spectroscopy and a differential scanning calorimetry study of the physical properties of N-acylphosphatidylethanolamines in aqueous dispersions. Chem. Phys. Lipids 46, 43-50.

AL-SALEH A., MARION D., GALLANT D.J. (1986) Microstructure of mealy and vitreous endosperms (*Triticum durum* L.) with special emphasis on location and polymorphic behavior of lipids. Food Microstruct. 5, 131-140.

BALLS A.K., Hale W.S. (1940) A sulphur bearing constituent of the petroleum ether extract of wheat flour (preliminary report). Cereal Chem. 17, 243-245.

BARNES (1983) Non-saponifiable lipids in cereals. In: Lipids in Cereal Technology (BARNES P.J., ed), Academic Press, New York, pp. 33-55.

BEKES F., SMIED I. (1981). Essay into the protein-lipid complexes of wheat flour soluble in petroleum ether. Acta.Alimentaria., 10, 229-253

BLOCHET J.E., KABOULOU A.; COMPOINT J.P.; MARION D. (1991) Amphiphilic proteins from wheat flour: Specific extraction, structure and lipid binding properties. In Gluten Proteins 1990; (BUSHUK W. and TKACHUK R., eds), A.A.C.C., St Paul, Minnesota, 1991, pp 314-325.

BLOCHET J.E., CHEVALIER C., FOREST E., PEBAY-PEYROULA E., GAUTIER M.-F., JOUDRIER P., PEZOLET M., MARION D. (1993) Complete amino acid sequence of puroindoline, a new basic and cystine rich protein with a unique tryptophan-rich domain, isolated from wheat endosperm by Triton X-114 phase partitioning. FEBS Lett. 329, 336-340.

BORDIER C. (1981) Phase separation of integral membrane proteins in Triton X114 solution. J. Biol. Chem. 25, 1604-1607.

BROOKER B.E. (1993) The stabilisation of air in foods containing fat- A review. Food Structure 12 115-122.

BRUIX M., JIMENEZ M.A., SANTORO J., GONZALES C., COLILLA F.J., MENDEZ E., RICO M. (1993) Solution structure of γ_1-H and γ_1-P thionins from barley and wheat endosperm determined by 1H NMR: a structural motif common to toxic arthropod proteins. Biochemistry 32, 715-724.

CARBONERO P., GARCIA-OLMEDO F., HERNANDEZ-LUCAS C. (1980) External association of hordothionin with protein bodies in mature barley. J.Agric. Food Chem. 28, 399-402.

CARLSON T., K. LARSSON, MIEZIS M, POOVARODOM S.(1979) Phase equilibria in the aqueous system of wheat gluten lipids and in the aqueous salt system of wheat lipids. Cereal Chem., 56, 417-419.

CARLSON T., LARSSON K., MIEZIS M., (1978), Phase equilibria and structures in aqueous system of wheat lipids. Cereal Chem., 55, 168-179.

CARRASCO L., VASQUEZ D., HERNANDEZ-LUCAS C., CARBONERO P., GARCIA-OLMEDO F. (1981) Thionins: plant peptides that modify membrane permeability in cultured mammalian cells. Eur. J. Biochem. 116, 185-188.

CHEN, J., DICKINSON, E., IVESON, G (1993) Interfacial interactions, comPetitive adsorption and emulsion stability Food Structure, 12, 135-146.

CHRISTIE W., MORISSON W.R. (1988) Separation of complex lipids of cereals by high performance liquid chromatography. J.Chromatogr. 456, 510-513.

CLARK D.C., COKE M., MACKIE A.R., PINDER A.C., WILSON D.R. (1990) Molecular diffusion and thickness measurements ôf protein stabilized thin liquid films. J. Colloid Interface Sci. 138 (1990) 207-218.

CLARK D.C., WILDE P.J., BERGINK-MARTENS D., KOKELAAR A., PRINS A. (1993) Surface dilational behaviour of aqueous solutions of β-lactoglobulin and Tween 20. In 'Food Colloids and Polymers: Structure and Dynamics', RSC Special Publication, pp 354-364.

CLARK D.C., WILDE P.J., MARION D. (1994) The effect of lipid binding protein on the foaming properties of beer containing lipid. J.Inst.Brew.100, 23-25.

CLARK D.C., COKE M., MACKIE A.R., PINDER A.C., WILSON, D.R. (1990) Molecular diffusion and thickness measurements of protein stabilized thin liquid films. J.Colloid Interf. Sci. 138, 207-218.

CLORE G.M., SUKUMURA D.K., GRONENBORN A.M., TEETER M.M., WHITLOW M., JONES B.L. (1987) Nuclear magnetic resonance study of the solution structure of α_1-purothionin. Sequential resonance assignment, secondary structure and low resolution tertiary structure. J. Mol. Biol 193, 571-578.

COKE M., WILDE P.J., RUSSELL E.J., CLARK,D.C. (1990) The influence of surface composition and molecular diffusion on the stability of foams formed from protein detergent mixtures. J. Colloid Interface Sci. 138, 489-504.

COLILLA F.J., ROCHER A., MENDEZ E. (1990) γ-purothionins: amino acid sequence of two polypeptides of a new family of thionins from wheat endosperm. FEBS Lett. 270, 191-194.

CROWE L.M., CROWE J.H. (1982) Hydratation dependent hexagonal phase lipid in a biological membrane. Arch. Biochem. Biophys 217, 582-587.

CULLIS P.R., DE KRUIJFF B. (1978) The polymorphic behavior of phosphatidylethanolamines of natural and synthetic origin. A ^{31}P NMR study. Biochim. Biophys. Acta 513, 31-42.

CULLIS P.R., DE KRUIJFF B. (1979) Lipid polymorphism and the functional role of lipids in biological membranes. Biochim. Biophys. Acta 559, 399-420.

CULLIS P.R., HOPE M.J., TILCOCK C.P.S. (1986) Lipid polymorphism and the role of lipid in membranes. Chem. Phys. Lipids 40, 127-144.
DESORMEAUX, A.; BLOCHET, J.E.; PEZOLET, M.; MARION D. Amino acid sequence of a non-specific wheat phospholipid binding protein and its conformation revealed by infra-red and Raman spectroscopy. Role of disulphide bridges and phospholipids in stabilizing the α-helix. Biochim. Biophys.Acta 1992, 1121, 137-152.
DICKINSON, E, WOSKETT, C.M. (1989) Competitive adsorption between proteins and small-molecule surfactants in food emulsions. In Food Colloids (BEE R.B, MINGINS J., RICHMOND P., eds), Royal Society of Chemistry Special Publication No.75, London, pp.74-96.
DUBREIL L., QUILLIEN L., COMPOINT J.P., MARION D. (1994) Variability, biosynthesis, degradation and location of wheat indolines and lipid transfer proteins. In "Wheat Kernel Proteins, Molecular and Functional Aspects" , CNR, Viterbo, pp 331-333.
FAUCON J.F., MELEARD P. (1993) Polymorphisme des lipides. Contraintes stériques et élastiques. In Les liposomes. Aspects technologiques, biologiques et pharmacologiques. (DELATTRE J., COUVREUR P., PUISIEUX F., PHILIPPOT J.R., SCHUBER F., eds), éditions INSERM, Paris, pp 7-42.
FISCHWICK M.J., WRIGHT A.J. (1980) Isolation and characterization of amyloplast envelope membranes from *Solanum tuberosum*. Phytochemistry 19, 55-59.
FRAZIER P.J. (1983) Lipid-protein interactions during dough development. In: Lipids in Cereal Technology (BARNES P.J., ed), Academic Press, New York, pp 189-212.
FRAZIER P.J., DANIELS N.W.R., RUSSEL-EGGIT P.W. (1981) Lipid-protein interaction during dough development. J. Sci. Food Agric. 32, 877-897.
FULLINGTON J.G. (1967) Interaction of phospholipid-metal complexes with water-soluble wheat protein. J.Lipid Res. 8, 609-614.
GALLIARD T.(1983) Enzymic degradation of cereal lipids. In: Lipids in Cereal Technology (BARNES P.J., ed), Academic Press, New York, pp 111-148.
GAN Z., ANGOLD R.E., WILLIAMS M.R., ELLIS P.R., VAUGHAN J.G., GALLIARD T. (1990) The microstructure and gas retention of bread dough. J. Cereal Sci., 12, 15-24.
GARCIA-MAROTO F., MARANA C., MENA M., GARCIA-OLMEDO F., CARBONERO P. (1990) Cloning of cDNA and chromosomal location of genes encoding the three types of subunits of the wheat tetrameric inhibitor of insect α-amylase. Plant Mol. Biol. 14, 845-853.

GAUTIER M.F., ALEMAN M.E., GUIRAO A., MARION D., JOUDRIER P. (1994) *Triticum aestivum* puroindolines, two basic cystine-rich seed proteins: cDNA sequence analysis and developmental gene expression. Plant Mol Biol. 25, 43-57

GENNIS R.B.(1986) Biomembranes-Molecular structure and function. Springer-Verlag, New-York.

GINCEL E., SIMORRE J.P., CAILLE A., MARION D., PTAK M., VOVELLE F. (1994) Three-dimensional structure in solution of a wheat lipid transfer protein from multidimensional 1H-NMR data- a new folding for lipid carriers. Eur.J.Biochem. 226, 413-422.

GOMAR J., PETIT M.C., SODANO P., SY D., MARION D., KADER J.C., VOVELLE F., PTAK M. (1996) Solution structure and lipid binding of a non-specific lipid transfer protein extracted from maize seeds. Protein Sci., in press.

GORDON-KAMM W.J., STEPONKUS P.L. (1984) Lamellar to hexagonal II phase transitions in the plasma membrane of isolated protoplasts after freeze-induced dehydration. Proc. Natl. Acad. Sci. U.S.A. 81, 6373-6377.

GREENBLATT G.A., BETTGE A.D., MORRIS C.F. (1994) Relationship between endosperm texture and the occurence of friabilin and bound polar lipids on wheat starch. Cereal Chem. 72, 172-176.

GREENWELL P., SCHOFIELD J.D. (1986) A starch granule protein associated with endosperm softness in wheat. Cereal Chem. 63, 379-380.

GROSSKREUTZ (1961) A lipoprotein model of wheat gluten structure. Cereal Chem. 38, 336-349.

HARGIN K.D., MORRISON W.R. (1980) The distribution of acyl lipids in the germ, aleurone, starch and non-starcch endosperm of four wheat flour varieties. J. Sci. Food Agric. 31, 877-888.

HARGREAVES J., POPINEAU Y., MARION D., LEFEVRE J., LE MESTE M. (1995) Gluten viscoelasticity is not lipid mediated. A rheological and molecular flexibility study on lipid and non prolamin protein depleted glutens. J.Agric.Food Chem. 43, 1170-1176.

HAWGOOD S., BENSON B.J., SCHILLING J., DAMM D., CLEMENTS J.A., WHITE R.T. (1987) Nucleotide and amino acid sequences of pulmonary surfactant protein SP18 and evidence for cooperation between SP18 and SP 28-36 in surfactant adsorption. Proc. Natl. Acad. Sci. U.S.A. 84, 66-70.

HAWGOOD S., CLEMENTS J.A. (1990) Pulmonary surfactant and its apoproteins J. Clin. Invest. 86, 1-6.

HEINEMANN B., ANDERSEN K.M., NIELSEN P.R., BECH L.M., POULSEN F.M. (1996) Structure in solution of a four-helix lipid binding protein. Protein Sci. 5, 13-23.

HERNANDEZ-LUCAS C., FERNANDE DE CALEYA R., CARBONERO P., GARCIA-OLMEDO F. (1977) Reconstitution of petroleum ether soluble wheat lipopurothionin by adding of DGDG to the chloroform soluble form. J.Agric. Food Chem. 25, 1287-1289.
HESS K., MAHL H. (1954) Elektronmikroskopische Beobachtungen an Mehl und Mehlpreparation von Weizen. Mikroskopie, 9,81.
HOSENEY R.C., FINNEY K.F., POMERANZ Y. (1970) Functional (breadmaking) and biochemical properties of wheat flour components.VI. Gliadin-lipid-glutenin interaction in wheat gluten. Cereal Chem. 47, 135-140.
HUSBAND F., WILDE P.J., MARION D., CLARK D.C. (1994) A comparison of the foaming and interfacial properties of two related lipid binding proteins from wheat in the presence of a competiting surfactant. In "Food Macromolecules and Colloids" (DICKINSON E. and LORIENT.D., eds), Royal Society of Chemistry, London, pp 285-296.
JOLLY, C.J., RAHMAN, S., KORTT, A.A. and HIGGINS, T.J.V. Characterization of the wheat Mr 15000 grain-softness protein and analysis of the relationship between its accumulation in the whole seed and grain softness. Theoretical and Applied Genetics 86 (1993) 589-597.
JONES B.L., MAK A.S. (1977). Amino acid sequences of two A-purothionins of hexaploid wheat. Cereal Chem., 16, 511-523.
KADER J.C. (1993) Lipid transport in plants. In Lipid metabolism in plants (MOORE T.S. Jr, ed.), CRC Press, Boca Ronta, Florida, pp 309-336.
KEE C.B. Ca^{2+}-dependent phospholipid- (and membrane) binding proteins. Biochemistry 27, 6645-6653.
KILLIAN J.A., SALEMINK I., DE PLANQUE M.R.R., LINDBLOM G., KOEPPE II, R.E., GREATHOUSE D.V. (1996) Induction of nonbilayer structure in diacylphosphatidylcholine model membranes by transmembrane a-helical peptides: importance of hydrophobic mismatch and proposed role of tryptophans. Biochemistry 35, 1037-1045.
KING R.J. (1984) Isolation and chemical composition of pulmonary surfactant. In Pulmonary Surfactant (ROBERTSON et al., eds), Elsevier, Amsterdam, pp 1-15.
LAFRANCE D., MARION D. PEZOLET M. (1990) Study of the structure of N-acyldipalmitoylphosphatidylethanollamines in aqueous dispersion by infrared and Raman spectroscopy. Biochemistry 29, 4592-4599.
LALCHEV Z.I., WILDE P.J., CLARK D.C. (1993) Surface diffusion in lipid films.1 Dependence of the diffusion coefficient on the lipid phase state, molecular length and charge J. Colloid Interface Sci., 167, 80-86.
LANDOLT-MATICORENA C., WILLIAMS K.A., DEBER C.M., REITHMEIER R.A.F. (1993) Non-random distribution of amino acids in the transmembrane segments of human type I single span membrane proteins. J.Mol. Biol 229, 602-608.

LARSSON K. (1986). Functionality of wheat lipids in relation to gluten gel formation. In Chemistry and Physics of Baking. (BLANSHARD J.M.V., FRAZIER J.P., GALLIARD T. Eds), Royal Society of Chemistry, Londres, p. 62-64.
LE ROUX C., MARION D., BIZOT H., GALLANT D. (1990). Thermotropic behaviour of coconut oil during dough mixing: evidence for a liquid/solid phase separation according to mixing temperature. Food Microstruct.9, 123-131.
LINDBLOM G., RILFORS L. (1989) Cubic phases and isotropic structures formed by membrane lipids: possible biological relevance. Biochim. Biophys. Acta. 988, 221-256.
LUZZATI V., (1968) X- ray diffraction studies of lipid-water systems. In: Biological Membranes. (CHAPMAN D., ed), Academic Press, New York, pp 71-223.
LUZZATI V., GULIK-KRZYWICKI T., TARDIEU A. (1968) Polymorphism of lecithins. Nature 218, 1031-1034.
LUZZATI V., HUSSON R. (1962) The structure of the liquid-crystalline phases of lipid-water systems. J.Cell Biol. 12, 207-219.
MACRITCHIE F.(1983) The role of lipids in baking. In: Lipids in Cereal Technology (BARNES P.J., ed), Academic Press, New York, pp 165-188.
MACRITCHIE F., GRAS P.W. (1973) The role of flour lipids in baking. Cereal Chem. 50, 292-302.
MADDEN T.D., CULLIS P.R.(1982) Stabilization of bilayer structure for unsaturated phosphatidylethanolamines by detergents. Biochim.Biophys. Acta 684, 149-153.
MAK A.S., JONES B.L. (1976). The amino acid sequence of wheat ß-purothionin. Can. J. Biochem. 54, 835-842.
MARION D., CLARK D.C. (1996). Wheat lipids and lipid binding proteins: structure and function, in press.
MARION D., DOUILLARD R., GANDEMER G. (1984) Separation of plant phosphoglycerides and galactosylglycerides by high performance liquid chromatography. In Structure, Function and Metabolism of Plant Lipids (SIEGENTHALER P.A. and EICHENBERGER W., eds), Elsevier, Amsterdam, pp 139-143.
MARION D., LE ROUX C., AKOKA S., TELLIER C., GALLANT D. (1987) Lipid-protein interactions in wheat gluten: a phosphorus magnetic resonance spectroscopy and freeze-fracture electron microscopy study. J. Cereal Sci. 5, 101-115.
MARION D., LE ROUX C., TELLIER C., AKOKA S., GALLANT D., GUEGUEN J., POPINEAU Y., COMPOINT J.P. (1989) Lipid-Protein interactions in wheat gluten: a renewal. In: Interactions in Protein Systems" (SCHWENKE K.D. et RAAB B.,eds), Springer Verlag, Berlin, p147-152 et p 363-373.

MARSH D. (1991) General features of phospholipid phase transitions. Chem. Phys. Lipids 57, 109-120.
MORRIS C.F., GREENBLATT G.A., BETTGE A.D., MALKAWI H.I. (1992) Isolation and characterization of multiple form of friabilin. J. Cereal Sci. 15, 143-149.
MORRISON W.R. (1988) Lipids. In: Wheat chemistry and Technology (POMERANZ Y., ed), AACC, St-Paul, Minnesota, vol.1, pp 373-439.
MURPHY D.J. (1990) Storage lipid bodies in plants and other organisms. Prog. Lipid Res. 29, 299-324.
NEUMANN G.M., CONDRON R., THOMAS I., POLYA G.M. (1994) Purification and sequencing of a family of wheat lipid transfer protein homologues phosphorylated by plant calcium-dependent protein kinase. Biochim. Biophys. Acta 1209, 183-190.
NICOLAS J., DRAPRON R. (1983) Lipoxygenase and some related enzymes in breadmaking. In: Lipids in Cereal Technology (BARNES P.J., ed), Academic Press, New York, pp 213-235.
NOTTER R.H. (1984) Surface chemistry of pulmonary surfactant: the role of individual components. In Pulmonary Surfactant (ROBERTSON B., VAN GOLDE L.M.G., BATENBURG J.J., eds), Elsevier, Amsterdam, pp 17-65.
OLCOTT H.S., MECHAM D.K. (1947) Characterization of wheat gluten. I. protein-lipid complex formation during doughing of flours. Lipoprotein nature of glutenin fraction. Cereal Chem. 24, 407-414.
PATTUS F., DESNUELLE P., VERGER R. (1978) Spreading of liposomes at the air-water interface. Biochim. Biophys. Acta 507, 62-70.
PEBAY-PEYROULA E., COHEN-ADDAD C., LEHMANN M.S., MARION D. (1992) Crystallographic data for the 9000 Dalton non-specific phospholipid transfer protein. J.Mol.Biol. 226, 563-564.
PELESE-SIEBENBOURG F., CAELLES C., KAFER J.C., DELSENY M., PUIGDOMENECH P. (1994) A pair of genes coding for lipid transfer proteins in sorghum vulgare. Gene 148, 305-308.
PETIT M.C., SODANO P., MARION D., PTAK M. (1994) Two-dimensional 1H-NMR studies of maize lipid transfer protein -sequence specific assignment and secondary structure. Eur.J.Biochem. 222, 1047-1054.
PRYDE J.G., PHILLIPS J.H., (1986). Fractionation of membrane proteins by temperature-induced phase. Biochem. J., 233, 525-533.
RAJAPAKSA D., ELIASSON A.C., LARSSON K. (1983) Bread baked from wheat-rice mixed flours using liquid-crystalline lipid phases in order to improve bread volume. J.Cereal Sci. 1, 53-61.
REDMAN D.G, EWART J.A.D.(1973) Characterisation of three wheat proteins found in chloroform-methanol extract. J.Sci Food Agric. 24, 629-636.
SARKER, D.K.,WILDE P.J., CLARK D.C., (1995) Competitive adsorption of L-α-lysophosphatidylcholine/.β-lactoglobulin mixtures at the interfaces of foams and foam lamellae Colloids and Surfaces B: Biointerfaces. 3, 349-356.

SCHIFFER M., CHANG C.H., STEVENS F.J. (1992) The functions of tryptophan residues in membrane proteins. Protein Engineer. 5, 213-214.
SCHINDLER, H. (1979) Exchange and interactions between lipid layers at the surface of a liposome solution. Biochim. Biophys. Acta 555, 316-336.
SEDDON J.M. (1987) Structure of the inverted hexagonal HII phase, and non-lamellar phase transitions of lipids. Biochim. Biophys. Acta 1031, 1-69.
SHIN D.H., LEE J.Y., QWANG K.Y., KIM K.K., SUH S.W. (1995) High resolution crystal structure of the non specific lipid transfer protein from maize seedlings. Structure 3, 189-199.
SIMON E.W. (1978) Membranes in dry and imbibiting seeds. In Dry biological systems (CROWE and CROWE, eds), Academic press, New York, pp 205-224.
SIMORRE J.P., CAILLE A. MARION D. MARION D., PTAK M. (1991) Two and three-dimensional 1H-NMR studies of a wheat phospholipid transfer protein. Sequential assignments and secondary structure. Biochemistry 30, 11600-11608
SINGER S.J., NICOLSON G.L. (1972) The fluid mosaic model of the structure of cell membranes. Science 175, 720-731.
SKRIVER K., LEAH R., MÜLLER-URI F., OLSEN F.L., MUNDY J. (1992) Structure and expression of the barley lipid transfer protein gene Ltp1. Plant Mol. Biol. 18, 585-589.
SMALL D. (1986) The physical chemistry of lipids from alkanes to phospholipids. Handbook of Lipid Research, Plenum Press, New York, vol. 4.
SMALLWOOD M., KEEN J.N., BOWLES D.J. (1990) Purification and partial sequence analysis of plant annexins. Biochem. J. 270, 157-161.
SORENSEN S.B., BECH L.M., MULDBJERG T.B., BREDDAM K. (1993) Barley lipid transfer protein 1 is involved in beer foam formation. MBAA Technical Quater. 30, 136-145.
STEVENS D.J. (1959). The contribution of the germ to the oil content of white flour. Cereal Chem., 36, 452-461.
SUBIRADE M., MARION D., PEZOLET M. (1993) Interaction of a non specific maize phospholipid transfer protein with unilamellar phospholipid vesicles : an infrared spectroscopy study. Proceedings of the 9th International Conference on Fourier Transform Spectroscopy, S.P.I.E. vol 2089, pp 346-347.
SUBIRADE M., SALESSE C., MARION D., PéZOLET M. (1995) Interaction of a non specific wheat lipid transfer protein with phospholipid monolayers imaged by fluorescence microscopy and studied by infrated spectroscopy. Biophys. J. 69, 974-978.
TATE M.W., EIKENBERRY E.F., TURNER D.C., SHYAMSUNDER E., GRUNER S.M. (1991) Nonbilayer phases of membrane lipids. Chem. Phys. Lipids 57, 147-164.

TCHANG F., THIS P., STIEFEL V., ARONDEL V., MORCH M.D., PAGES M., PUIGDOMENECH P., GRELLET F., DELSENY M., BOUILLON P., HUET J.C., GUERBETTE F., BEAUVAIS-CANTE F., DURANTON H., PERNOLLET J.C., KADER J.C. (1988) Phospholipid transfer protein: full length cDNA and amino acid sequence in maize. Amino acid sequence homologies between plant lipid transfer proteins. J. Biol. Chem. 263, 16849-16856.

TEETER M.M., MA X., RAO U., WHITLOW M. (1990) Crystal structure of a protein-toxin a1-purothionin at 2.5Å and a comparison with predicted models. Proteins: Struc.,Func.,Genet 8, 118-132.

TSUBOI S., SUGA T., TAKISHIMA K., MAMIYA G., MATSUI K., OZEKI Y., YAMADA M. (1991) Organ specific occurrence and expression of the isoforms of non specific lipid transfer protein in castor bean seedlings and molecular cloning of a full-length cDNA for a cotyledon specific isoform. J.Biochem (Tokyo) 110, 823-831.

VAIL W.I., STOLLERY J.G. (1979) Phase changes of cardiolipin vesicles mediated by divalent cations. Biochim. Biophys. Acta 551, 74-84.

VAN ECHTELD C.J.A., DE KRUIJFF B., VERKLEIJ A.J., LEUNISSEN-BIJVELT J., DE GIER J. (1982) Gramicidin induces the formation of non-bilayer structures in phosphatidylcholine dispersions in a fatty acid chain dependent way. Biochim. Biophys. Acta 692, 126-138.

WEISS M.S., ABELE U., WECKESSER J., WELTE W., SCHILTZ E., SCHULZ G.E. (1991) Molecular architecture and electrostatic properties of a bacterial porin. Science, 254, 1627-1630.

WHERLI H.P., POMERANZ Y. (1970) A note on the interaction between glycolipids and wheat flour macromolecules. Cereal Chem. 47, 160-166.

WILDE P.J., CLARK D.C., MARION D. (1993) The influence of competitive adsorption of a lysopalmitoyl phosphatidylcholine on the functional properties of puroindoline, a lipid binding protein isolated from wheat flour. J.Agric. Food Chem., 41, 1570-1576.

WIRTZ K.W.A., OP DEN KAMP J.A.F., ROELOFSEN B. (1986) Phosphatidylcholine transfer protein: properties and applications in membrane research. In Lipid-Protein Interactions (WATTS A. and DE PONT J.J.H.H.M., eds), Elsevier, Amsterdam, pp 221-265.

YU S-H., HARDING P., POSSMAYER F. (1984) Artificial pulmonary surfactant. Potential role for hexagonal HII phase in the formation of a surface active monolayer. Biochim. Biophys. Acta 776, 37-47.

ZAWISTOWSKA U., BEKES F., BUSHUK W. (1985) Gluten proteins with high affinity to flour lipids. Cereal Chem. 62, 284-289.

ZAWISTOWSKA U., BUSHUK W. (1986). Electrophoretic characterisation of low-molecular weight wheat protein of variable solubility. J. Sci. Food Agric.37, 409-417.

INDEX